# PLANET GRIEF

# PLANET GRIEF

## REDEFINING GRIEF FOR THE REAL WORLD

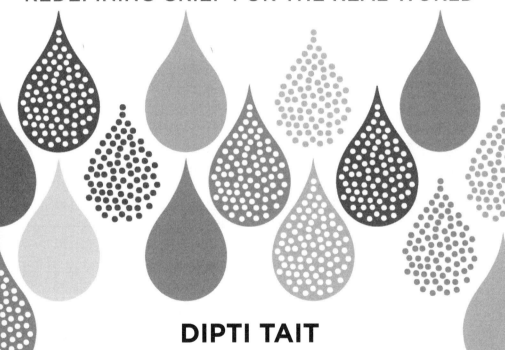

## DIPTI TAIT

FL☒NT

First published 2021

FLINT is an imprint of The History Press
97 St George's Place, Cheltenham,
Gloucestershire, GL50 3QB
www.flintbooks.co.uk

British Library Cataloguing in Publication Data.
A catalogue record for this book is available from the British Library.

ISBN 978 0 7509 9464 4

Typesetting and origination by The History Press
Printed and bound in Great Britain by TJ Books Limited, Padstow, Cornwall.

Trees for LYfe

# CONTENTS

# FOREWORD BY
# SHARRON DAVIES MBE

Not everyone can have a good friend like Dipti to listen to us when we need it, which we certainly all do from time to time, so it's great she's put some wise words into a book.

Grief is often our companion and through working with Dipti I've learnt you can make it something you get used to hanging out with. Wishing something wasn't so won't change what has already happened, sadly. We can only learn from mistakes, build support structures and become resilient. When I lost my mum, it was about learning to be grateful for the years I had and knowing there will be days when I'm more aware of the hole she leaves behind. I'm not sure that time repairs a broken heart, but time certainly enables us to live with our grief, whatever form it takes. I've always been a glass-half-full kind of person, but occasionally every one of us gets lost in a dark place. Sometimes we don't even realise we are grieving because we don't associate what we've lost with life-changing loss, but in reality our lives take different directions every day, acquiring new things and losing others. It's good to have Dipti's guidance on how we deal with the flux of life as best we can, particularly when we forget that the sun will always come up on another fresh day.

Sharron Davies MBE
June 2021

# FOREWORD BY PENNY POWER OBE

When we hear words such as vulnerability, self-esteem or confidence – human emotions that we all struggle with at some point in our lives – we all understand what they mean. We can relate to the words in our own way. These words might even flip your mind to a moment, a person, a need – they might even make you think of an expert who is defined by their wisdom.

This book is about grief, about Planet Grief, about all the ways that we grieve throughout our lives on Earth; after all, grief is a beautiful emotion that life gives us. So why is it that when we are confronted with the word 'grief', and we think of our relationship with it, many of us want to run in the opposite direction? We think of one thing – someone we loved has died.

Dipti taught me, and will teach you, about the diversity of grief, and that grief is essentially about loss. Loss is a life experience, in its many forms. Once you have read this book, you will change your relationship with loss – you will know when you are feeling loss, you will label it as 'grief', and you will give it the respect it deserves. Those moments when you are adjusting to change, when you feel uncomfortable, perhaps sad, and are required to dig deep inside yourself in order to move forwards – you will know that you are grieving, and that process is important to your future. Without grieving, we never find closure; fully accepting what has happened enables us to then find the courage and energy to see the new opportunities in the new space that has been created by what has gone.

I know loss in its many forms: I know the loss of loved ones, I know the loss of hope and I know the loss of a business. I know the loss of children running around my home, I know the loss of my dogs and I know the loss of our family home. I know the void these losses created, and I know the strategy of the 'stiff upper lip' of survival. I also know the damage of failing to grieve in the way that I should have.

We can all find hope, joy and a new energy to fill the space that a loss has created. There are mechanisms we can learn and adopt. Loss builds our resilience, it builds our empathy, and it does build new opportunities. This is the opportunity for you to learn to see the void and learn to understand the space that has opened up for you, and what it will mean for your future. In loss – in grief – there is transformation.

We should never just 'put up with the loss', but we should also never be defined by it. We can chart a positive way to find joy again, and enjoy this world we are stepping into. It will transform us, heal us and give us the strategies we need to live with all the inevitable losses we will have, big and small.

With love to you and wishing you courage and joy.

Penny Power OBE

June 2021

# ABOUT THE AUTHOR

In her early twenties, Dipti Tait started her working life at the BBC in London, and while there she found out her father only had months to live. He was diagnosed with leukaemia and died after only fourteen weeks.

Two years later, Dipti discovered she was having a baby and put her career in the media on hold to become a mum, and a year later had another baby. Becoming a full-time mother to two boys under 2, as well as giving up her beloved career in television and radio, and then moving away from her life and roots in London to relocate to the Cotswolds, made her realise that grief took many shapes and forms.

Her mother was soon after diagnosed with terminal cancer and died suddenly. Dipti was going through grief struggles as an only child, suffering with extreme loneliness, as she was also in the middle of a divorce and moving out of her family home with very little support.

This catalogue of devastating loss and change plunged her into a dark place. Parentless, and at rock bottom, she found a way to begin her life again, independently with two small boys.

Through this deeply troubled time, Dipti realised that grief was helping her to transform her life.

In 2016, she self-published her first book, *Good Grief*, as a way of coping with change and her losses, while she retrained as a solution-focused hypnotherapist to help others. She single-handedly built a successful online therapy business and to date has helped thousands of people cope with loss and grief, and build their mental and emotional wellbeing.

Now approaching her fifties, with her children grown into young adults about to fly the nest and her career taking her in a global direction, Dipti has built a meaningful life, and she has grief to thank for all of it.

Dipti passionately teaches her clients how to accept and be grateful for grief in life. She firmly believes that when we understand how to navigate and transition through grief, the transformation can be powerful and utterly life-changing.

Dipti is a professional public speaker and regularly contributes to BBC radio debates and has been a guest on ITV's *This Morning* and *Good Morning Britain*. See more about Dipti at diptitait.com.

# ACKNOWLEDGEMENTS

There are two special people whom I owe this book to. They gave me life and death: Ram and Manju Paul, my parents.

It's with both of you that I desperately want to share this book, but I can't. The paradox is that this book would not even exist if you *were* still here. How ironic. Instead of sadness, I fill my heart with gratitude and I thank you both from the bottom of my heart for everything you did for me and all the sacrifices you made for me. I love you deeply.

Jo de Vries, my incredibly kind friend and commissioning editor – as soon as you came into my life, this book began to emerge in a silent but powerful way. Without your encouragement, belief and gently forceful nudging, it would never have materialised. Thank you so much for being there for me – at all times, especially in this tough year of home-schooling.

The wonderful team at Flint Books – Laura Perehinec, publishing director, Chrissy McMorris, editorial manager, and Cynthia Hamilton, head of PR & marketing. Thank you so much for believing in my idea, and taking my words and transforming them into this beautiful offering. Your dedication and commitment to publishing shine through, and I am so blessed and grateful to be working with you all. Thank you so much to Katie Beard, head of design. The cover and design of this book are outstanding and capture the bright spirit and light-hearted essence of my words beautifully.

Thank you to Katie Read of Read Media, for listening to my ideas and musings, and for helping me share my message with the world. I am so grateful you are by my side – this is so reassuring.

Grief is such a delicate subject and I am grateful for the raw and beautiful conversations I had with all of my incredible contributors – you were brave and courageous to share your stories of loss, change and grief with me. Your impactful stories have enriched and infused these pages and will now help so many people. I would like to thank each of you *so much* for being honest and trusting me – Sharron, Penny, Elaine, Andy, Melanie, Becky, Maleena, Faraz, Jo, Charley, Sally, Pauline, Anna, Krishan, Liv and Ezra.

Thank you so much to the incredibly knowledgeable Dr Lynda Shaw, neuroscientist, for confirming so much of my own understanding about mental and emotional health with neuroscientific backup. I am so grateful for your time and patience. Your easy-to-understand explanations of how our brains and bodies work together in grief were extremely impactful and hugely engaging.

Thank you to James Bore, cyber security expert, for sharing some illuminating findings and comparisons between technology and how our own minds work. Talking to you was fascinating, and I thank you for your time and expertise.

Siân Storey, celebrant, it was so very lovely to connect with you and for you to give me such an interesting perspective on loss and grief from your professional experience. Thank you so much for your contribution.

Dearest Penny, you are my mentor, my friend and my business mum. Your interview made me cry, and I want you to know that your experience of grief and loss has been a confirmation of how we can turn grief into fuel, and this has been an amazing realisation. Thank you so much for writing my heartfelt foreword as well – that also made me cry!

My wonderful friend Sharron – your stories of grief are so powerful, and I know they will help many others who face similar challenges in life. Thank you for your constant unconditional kindness and support, and thank you for writing such a beautiful and thoughtful foreword to begin this book.

Thank you to *all* my clients – past, present and future. Without every single one of you, I would not be able to begin to understand the paradox of the simultaneous simplicity and complexity of the human mind – how delicate it can be, and how powerful it is. You teach me about myself every day and hold up the mirror for me to learn about how similarly and differently we as human beings behave and belong on the planet.

David Newton, you were my first mentor into the human mind. Thank you for giving me the gift of hypnotherapy. Neil Crofts, Robert Dilts, James Tripp, Adam Eason, Ian McDermott, Tim Hallbom – learning from you has shaped my understanding of therapy, coaching, the mind and our emotions, and I can't thank you all enough.

Thank you to my gorgeous boys. Krishan, my eldest, and Jacob, my youngest, for making me burst with pride because of who you both are. You are both my ultimate driving force and my reason to stay strong and always be the change that I want to see in the world. Thank you for being my constant inspiration. You are both perfect in my eyes, and always will be.

Toby. Without you, I would not be where I am today. You are the force beside me that keeps me going. You are my soul mate, forever. Thank you for loving me and supporting me with everything I do. You are my rock, my hero and my best friend. I love you with every cell of my body and every molecule of my mind. (Oh, and PS, thank you for buying me noise-cancelling AirPods so I could shut out the loud and chaotic pandemic-stricken world and fully concentrate on writing this book in the middle of a global crisis.)

Sadly, while writing this book, I have personally known people who have left this planet. Saima Thompson, Amal Ray, Scott McCormick, Neil Hadaway and Clare Dunkel. My thoughts and heartfelt wishes extend out to all of their families and friends. May you all rest in peace.

I dedicate this book to all of you who have faced loss and grief because of COVID-19. I am so deeply sorry, and I hope this book reminds you that you are not alone, and it's okay not to be okay.

COVID-19, you taught me so much more about our planet in deep grief. I don't feel 'thank you' is the right sentiment. But I do acknowledge the immense power you have brought into the planet with your arrival, as we all now begin to navigate into the 'new different'.

Finally, I acknowledge *grief*. Without grief, this book would not be in your hands.

Diddima,
grandmother, mentor and guru,
I wrote this book in honour of you.
Namasté.

# AUTHOR'S NOTE

Storytelling is a powerful way of connecting us with ideas and concepts, so throughout this book I've used stories to illustrate the ways in which grief can impact our lives and how we can learn to live more comfortably with that. Many of these stories are my own (and in some cases names and the detail of events have been changed for anonymity), while others have generously been told to me by clients or interviewees, some of whom have been anonymised to protect their privacy, and others are fictionalised case studies based on decades of observing how people respond to situations of loss and change. Where a story is fictionalised, I have made that clear beforehand and any similarities to real-life situations or individuals are coincidental and unintentional. It is also important to remember that although these stories are offered as a way of illustrating particular themes and ideas, they are not intended to be representative of age, gender, culture, ethnicity, religious and occupational experiences, which will be unique for everyone. Also, approaches that have worked for one person are not necessarily suited to every individual or circumstance, so it is vital that you find the right support and way forward for you.

If any of the stories in this book trigger issues for you, please be kind to yourself and do seek professional help. I've listed some sources of support in the back of the book, but this list is by no means exhaustive and the inclusion of these resources does not imply an endorsement of them – it is important that you find the right support for you with professional guidance.

# INTRODUCTION

'I'm not afraid of death anymore, and I think a lot of the struggle
we have in life comes from a deep, deep fear of death.'

*Naval Ravikant,*
*investor, entrepreneur and co-founder of AngelList*

I have a deep question for you to ponder ...

> If everybody you knew, everything you owned and all that you
> experienced in life had a 'shelf life' or 'end date' attached to it –
> which meant you knew exactly when they would leave your life
> – would you interact with them any differently?

What about your own life?

We know our 'start date' – the day we are born, the day we celebrate every year – as our birthday.

But what if you also knew your 'end date' – your death day?

This would also mean that you knew when your friendships would end, relationships would finish or careers would come to a stop. You would know when your parents were going to die and all your friends and family too.

You would also be able to know about when big things were going to happen to you and un-happen to you.

Perhaps you would have already been prepared for the losses that we have faced from COVID-19 coming into our lives in 2020?

If we were to know about these endings in advance, would this mean we would feel okay about them when they happened? Or would we live our lives very differently?

The biggest question for me is: would this prior knowledge of things leaving us, ending for us and dying on us be the end of grief? I don't think so.

A lot of human beings fear death, but there are some humans, like Naval Ravikant, who understand death and live life as fully as they can because they completely understand that there is a small window of opportunity that we all have on this planet when we are alive.

This window of opportunity for most human beings is between zero and a hundred years. Some human beings even live across two centuries, and that is basically our limit on the planet.

As I write this, I'm a month away from turning 48. If I live to around 81 years old – which is the average life expectancy for human beings living in the UK today – I have just over 33 years, or 396 months, or 1,716 weeks, or 12,045 days left on this planet. That doesn't seem like many weeks or days when you look at it like that, does it?

That would mean that my potential 'best before date' will be some time in 2054. In fact, now that I have just worked out a potential 'BBD' for myself, it does feel a bit more *real*. It also makes me feel like I can't waste any time.

I suddenly feel as if I have an urgent mission to get close to 2054 – when I will be able to look back and feel so happy with everything I achieved and all that I did in my life, the people I have met, the experiences I have had and the love I have felt.

I want to be able to leave this planet with a huge smile on my face and a wealth of happiness in my heart.

I may not know exactly when I will die, but I do know I will, and I do know that there will be people grieving my departure; even if they know it's coming, they will still grieve.

We are all creatures of grief, because we have the capacity to form attachments and bonds. Once these attachments and bonds are broken, we feel the loss – this is grief.

If we can learn to overcome the fear of grief, we can also overcome the fear of loss and, ultimately, the fear of death.

Grief is part of life's struggle, and our struggle with the fear of death can often mean we develop a fear of life.

My hope for all of us is that we learn how to transform all the fears that we may develop in life into fuel and power.

Humans grieve, animals grieve, bees grieve. Our planet has an intelligent way of healing itself (if human beings don't get in the way).

If we get out of our own way, we can tap into a healing, transformational intelligence within ourselves which I believe is our very own subconscious mind. We can turn our own grief into fuel and use this fuel to power our lives.

My hope is that this book will show you how.

# DEATH: LET'S GET IT OVER WITH

Most of us don't like thinking about dying nor enjoy planning for our own departures. We tend to do our best to avoid the death talk, push it to one side and change the subject very quickly, because it makes us feel deeply uncomfortable, plus most of the time we're just too busy getting on with our lives.

We work, we travel, we raise families, we exercise, we shop, we watch movies, we entertain ourselves, we socialise, we eat and drink, and we tick along quite nicely, thank you very much. Days and weeks merge into each other, the months roll on by and the years pass. Until one day, something happens, and we are stopped dead in our tracks as we come face to face with the inevitable end.

This rude interruption to our status quo means that we are forced to think about what has changed and to confront something we had hoped to continue avoiding or denying altogether.

Death isn't something that passes by silently and quickly without a trace: it's a biggie, like a tornado, that sweeps you off your feet and you have no idea *when*, *if* or even *how* you will ever land.

When you do land, it's more like a crash landing in a barren landscape that feels unfamiliar, rearranged and empty of the truths you have clung to for so long. You look around and there is nothing to hold on to for comfort; you call out and nobody replies. You attempt to feel your way around this unknown territory and it just seems surreal and scary – a bit like when you get up in the night and it's pitch black. You try to navigate the room you know so well and it suddenly feels entirely alien, and potentially threatening – who knows what's lurking out there (most likely the inevitable trip hazards of the socks, shoes and pants you left on the floor last night!). You then quietly realise that your eyes need to adjust to a brand-new world; it's like a different planet – a planet that only has you on it, you and death. This is when you get right up close and personal with death and it can either become a silent companion or, if you are not very careful, your worst enemy.

For many of my clients when they first come to see me, they will talk about the fear – the fear that comes with death, the fear of abandonment and abandoning, and the fear of death itself.

Death is a leveller. It does not discriminate or isolate. It connects our planet together in a beautifully exquisite way. It helps restore balance and order. It kills off everyone eventually, despite their size, their status and their ego. If there wasn't death, there could not be life. But this doesn't make us less scared of it, does it? It's so final, isn't it? It is game over. The last curtain call. Perhaps that's why for many of us talking about death is the ultimate uncomfortable conversation.

So, why am I talking about it at the beginning of this book? Because if we begin with what makes us uncomfortable and learn to lean into that discomfort, bit by bit, the discomfort feels doable.

Facing your discomfort and taking a good look at your anxieties and fears in a safe space and in a sensible way, where you feel well supported, can help you see through the things that once held you back. Once you begin to see through the layers of your own discomfort, you discover that there is some sort of hope there. There is another version of life that exists beyond your discomfort and you may find yourself carefully edging closer to a new reality of acceptance and curiosity.

Embrace discomfort, says Farrah Storr, editor-in-chief of *Cosmopolitan* and author of *The Discomfort Zone*. Farrah argues that 'Discomfort might just be the start of something wonderful ... You just have to take the first step out of your comfort zone and into your discomfort zone to feel those rewards.' If you can do that, the taut grip of your discomfort zone will loosen and your comfort zone will begin to grow larger in its place, providing you with a solid foundation for personal growth, internal development and harnessing the power of your mind.

What is rewarding about being uncomfortable, you may be sitting there and thinking ...

Well, if I tell you that it's the up-close-and-personal death experiences I have faced in my past that have rewarded me with incredible strengths and superpowers that I didn't even know I had back then, which I know for sure I have now – then doesn't that feel worthy of enduring a little unease on the way?

♦♦♦

# MY STORY

It was Christmas Eve 2011. I sat for hours in the darkness, twenty-five days after losing my mother to liver cancer, and ten days after losing my beloved grandmother, my Diddima, whom I was exceedingly close to – even more so than my own parents – to dementia. I had already lost my father to leukaemia twelve years before and, being an only child, I felt desperately alone. Losing my mum meant I was now an orphan. I was an orphan in the middle of a divorce.

My fifteen-year marriage had ended, and I had moved out of my large Cotswold family home into a tiny two-bedroom apartment with my two small children.

Moving out of my marriage was another whole different process of grief. I had to very quickly thicken my skin. People who I had considered my real friends judged me. Some of my family, through their own shock and understandable disbelief, could not comprehend what was happening and for a short time pushed me away. I was not popular.

Being from Indian heritage, divorce is very much frowned upon. I am the only person in my family to wear the divorce badge because it is simply not the 'done' thing. Even though, as I write this, over a decade has passed and my family are all talking to me again and life is lovely ... at that time I was very much cast out and my decisions were not supported. I was alone, and I had to deal with that.

In my state of aloneness, I realised something very poignant. I was *really* alone: not lonely in a crowded room, but actually on my own, dealing with my life with nobody to ask for help or advice. Because of this sense of real aloneness, I began to grow another strength on top of my thickened skin. A strength in my own truth.

Although there was a part of me that had some serious strength and courage, there was also an equally shaky and vulnerable part of me, now fully exposed. My mental health by this point was in an erratically vulnerable state. I had no idea that Orphan Grief was a thing until I experienced it first-hand.

Orphan Grief is a very unusual type of loss. It happens to us when we feel ill-equipped to deal with life as an adult with full agency and control over our lives. When we have our parents or similar guardians, who take care of us and take on the role of teacher and mentor, we as children and young adults learn to reference our existence and actions through our parental guidance.

There comes a stage when we also rebel against the guidance and this is a normal progression of growing up. If we lose parental guidance during our transition into adulthood – like I did, losing my dominant parent (my dad) in my early twenties – we experience Orphan Grief.

My dad was the one I went to for advice, and he was the one who made all the decisions and choices regarding my upbringing, so when he died, I suddenly didn't have a template of how to run my own life. Even though I craved independence and the agency and authority to be my own person, when I actually got it I felt scared and alone. I had no map, no compass, no direction. I was suddenly very lost.

I felt like I was drowning in deep grief-related anxiety and had obsessive fears about everything. My usual optimism and upbeat positivity were knocked out of me and I fell into a deep, dark hole of hopelessness.

I became withdrawn, didn't have a job, didn't want to socialise, and my mind was filled with irrational and anxious thoughts that kept me awake at night for weeks at a time.

Then, when my mother passed away, I was plunged into full Orphan Grief. At this time, my boys were 8 and 9 years old. They never knew what a state I was in. Like an unmarried Stepford Wife, I cooked as normal, I cleaned, I washed, ironed uniforms; I made the Easter hats with them, I went to the concerts, the plays, I helped at the fêtes, I helped with homework (if I was asked), I helped them sell their toys at a made-up shop outside our tiny flat; and we baked cakes and did fun chemistry experiments that involved fake blood and slime. We read stories together at bedtime and it was all good. I was a fun mum, I think, and more importantly – to them – I appeared fine. I felt strongly at the time that I needed to make sure that the transition of one home into two homes was as smooth and as normalised as it could possibly be.

So, I kept my emotions contained and I maintained my composure while they were with me. In some ways, I found this to be a very helpful process, because within the moments of holding myself together, bits of myself indeed seemed to be gluing themselves back together.

But, looking back, I was not fine. I was far from fine. I was smiling on the outside and crumbling on the inside. I felt like, if I let that carry on, I would stand up one day and disintegrate because there was nothing solid left inside me.

I maintained zombie-school-run-mum mode. I dutifully took the boys to school each morning and, after dropping them off and saying the obligatory 'hello, how are you' to the other parents and their teachers with a really convincing smile firmly planted on my face (my drama school days came in handy), I used to walk back to the car in tears – tears I felt I couldn't ever cry in front of the boys. I then got back into my little car and just sat there. I stared into space for seven hours – not moving, just sitting in that car, in complete silence, until it was home time again.

I repeated this weird routine for quite some time.

Luckily, the boys never asked me what I did every day, because I don't think my brain was active enough to come up with a pretend 'interesting' day. In fact, when I asked them what they did that day, they always had three standard replies: 'Nothing', 'I don't know' or 'I can't remember'. Those answers would have had to be my answers too.

My mind always seemed to be filled with noise and fog, and although there was so much noise, everything was also strangely muffled. It felt like my brain had been encased in layers and layers of bubble wrap and the bubbles were constantly popping and then refilling themselves with some kind of heavy gas – like sulphur hexafluoride (the gas that deepens the voice) – and this was weighing down my negatively charged mind.

I felt terribly gripped by a fear of the unknown and I couldn't visualise a positive future for myself. I felt alone, isolated and stuck. I realised I was deeply depressed.

Life had hit rock bottom.

One dark evening, shivering on a cold bathroom floor, as my dried-up tears started to sting life back into my face, I sat staring into a void of nothingness. I knew I had to do something before I was destroyed by my own thought-torture.

I chose to live, but the thoughts of meeting death by my own hand were very attractive, and at that time it was very difficult for me to push them away.

At night, when I got into bed, I would think the same catastrophic thoughts as I did every night – formulating a genius way to get out of this world cleanly, and wishing that I could find a clever way for it to look natural or like a freak accident. My night-time mind was constantly whirling with dark psychopathic plots that could have hatched from the mind of any best-selling thriller or horror writer.

Every night, I could feel my thoughts transforming themselves into a poison that would flood my body with tormented tension, silent stress and overwhelming panic – meaning I would always lie there, wide awake with adrenaline pumping through my system, keeping me on high alert until the small hours of the morning.

One night, my mind started racing in the usual way – it was becoming a habit – when suddenly, out of nowhere, everything went totally quiet. It was as though the cacophony in my head just cut out.

I was unusually calm, and in that stillness, I felt like I had tapped into a hidden resource that I never knew I had, and the only way I can describe it is as a strong urge to be okay, and an even stronger urge to survive.

I finally learnt to stop resisting those thoughts of ending my life, but instead, I found a new way to think about them. This may sound slightly weird, but believe me, it really helped. If a thought of ending my life came into my mind, I embraced it and mentally spoke to the thought. This is what I said: 'Hello. I see you, and I hear you. I know you feel real and you want my attention. I know you are giving me a choice to get out of this situation I am in right now. Thank you for giving me this choice.'

The strangest thing started to happen when I repeated this over and over. It was like the thought started to lose its power over me. I could breathe easier and I felt somehow internally held, heard and healed.

The suicidal thoughts were much like other thoughts now – I could allow them to pass through in the sky of my awareness; just like a little grey cloud, they squeezed out a few raindrops and then evaporated and disappeared. However, it is also very important that, if you do experience thoughts like these, you seek professional help, as it can often be impossible to conquer them on your own.

With these words implanted in my mind, I started to see the blue skies again and found the proverbial silver lining, and bit by bit I somehow slowly and painstakingly figured out what I wanted to do with my life.

There are really only ever two options: live or die.

I chose the first option – that is why you have this book in your hands.

At that time, my life felt like a gigantic box of jigsaw puzzle pieces had been emptied and scattered all over the floor and I had no idea how to put things back together again.

I realised I had to start slowly, piece by piece, and so I began to gain clarity. If I ever caught myself thinking about completing the whole puzzle, I would go backwards and feel like I had lost control, and this loss of control was overwhelming – way too overwhelming for me to even contemplate – so I had to start breaking my life down into manageable pieces again.

The first task I gave myself was to look for the corners in my life and then, after that, slowly think about building the frame.

The metaphorical corners of my life's puzzle at that time were: top right, Krishan, my son – he was 9 years old and such a bright boy, filled with intensity and a thirst for knowledge. His constant intellectual commentary of life made me laugh out loud, think deeply and feel so proud that I had such a wise conversationalist for a child.

Jacob, my other son – sat top left – such an enthusiastic and carefree kid; he was an excitable 8-year-old who saw life through the lens of curiosity and compassion. Everything was a fun experiment and his deeply honest joy for life brought tears to my eyes. He was lively, and there was never a dull moment with Jacob around.

Of course, they both needed me. I was their mother, their rock, their world and, of course, there was no way I could let them down.

This was a good start.

The memory of my grandma sat nicely bottom left. Diddima, I called her. She was always a cornerstone of my world and gave me lots of motivation while I was growing up with her dynamic energy and magical zest for life. She was fierce, though – a tough cookie – and she was small, but I would go so far as to bet that even Lennox Lewis would jump out of the ring if he was ever faced with Diddima. She could knock you out with one word that woman: she was a powerhouse of pure force.

The final corner piece for me – bottom right – was the trusted voice of my father. His gentle but admonishing voice was stamped firmly inside my mind as a source of truth and an audible mental reminder of the stuff I am made from. His voice underpinned my faith. His voice helped me get up and dust myself off and just keep going.

'Now that you have knocked the corner off the car, you can just get back in the driving seat and drive us to the scrapyard to get another indicator.' Those were his exact words when, only a week after passing my driving test, as I was turning into a driveway, I took the corner badly and shaved the indicator off on a low wall. I quickly got out of the car, totally mortified, and just stood there, immobile, looking at the damage and shaking with fear and sobbing with disbelief. Dad got out, came over to join me, looked at the damage and simply shook his head, then he punched me on the arm and said, 'It's okay. It's fixable. Everything is fixable, with the right parts.'

I hear that sentence every time I think about the bottom right corner of my puzzle, and it reminds me that if I ever cannot find the parts I need to get through life successfully, the scrapyard is a good place to go and, in amongst all that junk, there is this beautiful chaotic organisation!

The next part of the puzzle was to start connecting my corners with some kind of framework. The metaphorical idea of the frame helped me find some stability in my own emotional whirlwind. I used the frame to hold on to for support while I steadied myself and picked myself up from the deeply unhopeful and seriously unhelpful place I had found myself in.

While desperately clinging on to the frame, I felt like I was still standing on unsteady ground. But, after a little while, what started to happen is that I regained some balance, and the deaths and loss that had initially shaken my world started to solidify in my reality, and I felt them becoming a steady platform for me to stand on and to start rebuilding my life on again.

Part of my 'rebuild' was finding Toby. Toby came into my life as my mum was dying. While deep in grief, it was an interesting time to start a brand-new relationship, but a decade later we are still together, so that's a good sign! Of course, the beginning years of our relationship were hit and miss. We did break up once a month, and get back together; it was quite sketchy, but, weirdly, as soon as we started living together, we became solid. Toby is genuinely not just my rock, but my whole planet. He grounds me, supports me and helps me grow, and very skilfully keeps me centred. Toby and I had started to 'go out' with each other just before my mum passed away, and he could see that I was not in a good place at all. He suggested I asked Neil Crofts, co-founder of Holos Change, writer, coach and business consultant, for some help. I was reluctant, but Toby arranged it for me, and the next day Neil and I had an early Friday morning ninety-minute session booked over Skype.

Neil is like a real-life wizard and genius rolled into one human form. Neil shared with me a couple of pieces of his work. He sent me his utterly fascinating ebook *Seven Stages of Authenticity* to read, which totally blew my mind open, and he also took me through his extremely powerful 'Life Purpose Exercise'. His Life Purpose Exercise is very simple, but brings about deeply profound changes for people who are standing on the edge of a precipice – like I was. I fell into his work and his work caught me, supported me and sprung me back into life.

After going through these exercises, something different started to happen in my body and in my mind. I was getting rebuilt, from the inside. It felt like I was being reconsolidated and glued back together. I was getting the biggest system upgrade my nervous system could sustain and I was also in full operational defrag. Everything I didn't need was being destroyed, deleted or was being reused in a better way. I was re-emerging with a faster, more powerful operating system. I had, by the end of the weekend – a bit like Doctor Who – been completely regenerated.

I woke up on Monday morning and took the boys to school. This time, after the drop-off, I got in the car and drove back home. I wanted to do something very significant; I wanted to make a phone call to The Clifton Practice. This phone call was the start of my life's purpose, and it's a journey that has brought me here to writing this book.

By the end of that phone call, I was booked on to the ten-month hypnotherapy training programme with CPHT in Bristol. I started this new chapter of my life only two weeks after my coaching session with Neil Crofts, and it was the beginning of 2012.

In *Seven Stages of Authenticity*, Neil talks about life being a staged roadmap, but not a traditional linear map, rather one formed of concentric circles, like those found in a tree. At the centre of the circle is a dark hole, and as the circles get bigger, they evolve ... It's an amazing concept and, as I am a very visual and metaphorical learner, this appealed to me instantly, and it quite honestly changed my life.

My grief had forced me to stand in the very centre of my dark hole and realise that the darkness could easily engulf me, but with a little bit of awareness, I was able to shine some light on the darkness, which started to feel like pinpricks of brilliance punctuating the sorrow and hopelessness.

I like to think that, by now, I have managed to establish a really nice relationship with grief. I feel very thankful to grief for coming into my life and, if it wasn't for death, I would have never felt grief as much as I did, and would not be who I am today.

When I say that some people get angry with me because they don't understand what I mean. I have to explain that I'm not saying I am thankful that I've lost my wonderful parents and my beloved Diddima. Not at all. I would give everything to have them back in my world. But – and this is the bit I really need to highlight so I'll write it in bold – **If I hadn't experienced their deaths, I would never have met grief so intensely, and it's _that_ intensity that has made me the person I am today, and for _that_ I am grateful.**

I think grief shook me so much that it literally woke me up to a new reality. This sounds dramatic, but it's absolutely true. If my parents saw me now, they would not know who I was. I am not the same daughter they left behind.

I am not anybody's daughter any more, and that used to be the scariest feeling ever. Susan Jeffers coined the phrase 'feel the fear and do it anyway' in her 1987 book of the same name, and that has become my mantra as I choose to walk into that cold fear and step into the power that grief wants to offer. If it wasn't for grief waiting for me, I wouldn't be right here, where I am today.

I am now hope-full. (No, that's not a typo – I'm full of hope.)

So, my own story shows that death is monstrous and magnificent. Death is the elephant in the room, it's the huge pile of dust in the corner, it's the massive lump under the rug. It's the fact of life that we all wish was fiction and only happened to other people in faraway places, or in stories and films.

We don't like admitting death is real and it can happen to any of us, at any time. Admitting this means we have no control over death and when we feel out of control we often feel fear, anxiety and complete dread.

So, it's no wonder we don't want to think about it very much or talk about it a lot while we are alive and well.

But, in reality – our reality, on this beautiful planet we call Earth – death does exist. It will come for all of us. We cannot escape or stop it. We can run from it and prolong it, but eventually it visits us all. It also comes for our loved ones, our friends, our colleagues and our pets. There is no one it misses.

So, perhaps we have to stop thinking of death as a thing that exists in a vacuum – a monster under the bed – and join the idea of death up with life. We can do this by learning to accept loss and focus on finding happiness from the vantage point of sadness, finding stillness in the noise and harnessing the power of the storm whipping up inside our emotional worlds as we hold on to our own centre of gravity.

Boxing legend Mike Tyson views death as if it's his opponent in the ring: he gives it respect and leans into his passion, his purpose, and just makes sure his life packs a punch the best he can. Tyson recognised that every time he got in that ring he had to accept death was always an option. In one of his podcast episodes, he says, 'You know, we mustn't be afraid of death. We must look at death as being glorious.'

Similarly, if we all worked out how to accept that death is an option, we might become more at ease with it. As we embrace death more and more, and fully accept that, in the fight of life, death will always win, we may find some comfort there. We might learn to see death in a new way. We wouldn't die if we had never lived in the first place, so in order to live, we must learn to accept death.

Ding-ding.

## OUR BRAINS, OUR BODIES AND OUR GRIEF

Before we go any further in this book, it's really important to say that just because we started with death, that is far from the whole story of grief. *Grief is the stuff of life*. As we are about to explore, it runs through all our lived experiences, but in order to fully comprehend this, we need to understand some key things about what happens in our brains and bodies when we grieve. In my research for this book, I interviewed the cognitive neuroscientist Dr Lynda Shaw, who agreed that grief doesn't just occur when someone dies:

Grief can happen because of basically a sense of loss. And that loss can be bereavement, but it could be just due to change, for example if you're a child going to a new school and you feel a sense of loss for your old friends, or if there's been an illness, or unemployment, or moving house – all of those huge things can cause grief. So, it's actually that sense of loss that is a big issue. And with that sense of loss comes with it sometimes a feeling of loss of control. And when we feel that we are out of control, that can lead to depression, anxiety disorder, and other mental and physical illnesses.

All of these moments of change and flux in our lives can trigger grief, but grief in itself is not a single emotion – it is an experience. In fact, I often talk about grief's 'suitcase' of emotions, which I'll unpack later in the Global Grief chapter, but as Dr Shaw says, grief encompasses:

disbelief, separation, anxiety, all sorts of incredibly stressful feelings. And if you are grieving in that way, you're going to be experiencing a complete dance of your chemicals and hormones – they're going to be raging around all over the place. And that is when you can have specific symptoms like sleep loss, or eating too much or not eating enough. You can also have a very high level of fatigue and mental fatigue, that would come with problems such as memory issues, potential problems with concentration, and anxiety.

This is largely because when we grieve, as Dr Shaw points out, our brains and bodies are being flooded with a cocktail of chemicals, which can trigger a chronic level of stress if not addressed. In terms of what's going on inside of us when we grieve, she explains:

An area of the brain that is affected by grief is the prefrontal cortex and the frontal lobe, which govern our ability to find meaning, to express ourselves properly and our ability to have self-control. Therefore, in scenarios of grief, loss or trauma, all of these things will be compromised.

We have two biological pathways for stress, which complement one another. The first one is the sympathetic-adreno-medulla (SAM) axis. This is acute and is part of the sympathetic nervous system that activates the adrenal medulla, which releases adrenaline. So, the heart rate increases, your blood pressure goes up, you get a huge burst of energy because it is basically the fight or flight response – that's the body's initial response to a stress event. Once you've got over the shock and disbelief. This is tolerable, it's absolutely fine in the short-term.

However, there is also a second biological pathway in play, and that is the hypothalamus-pituitary-adrenal (HPA) axis, which is slower to respond. Of course, we're talking about the brain here, so it's not that slow, but it's slower. Basically, this is triggered by signals from the hypothalamus to the pituitary gland which leads to the release of cortisol from the adrenal cortex. Cortisol in small secretions is actually fine, in fact it's great for survival. It enhances the brain's use of glucose, which is the main source of energy for the brain. But then in the chronic situation of grief, that's when it gets toxic, because we are not meant to have high secretions of cortisol over a long period of time.

Essentially, if we are not careful, we can be 'held to ransom' by these secretions and grief can become a chronic illness. However, it doesn't have to be this way; by reaching a deeper understanding of the grief we experience in everyday life, we can apply our knowledge of the brain to help us deal with it. Later in this book we'll look at the different parts of the brain and how they influence our emotional responses, but Dr Shaw goes on to say:

You can't just make these [emotions] disappear. Because that's not how it works. They are part of your make-up now, the experience of that bereavement, the experience of the loss, for many people this stays with you. But when you are in a very high state of emotional arousal, you are not thinking clearly because emotion will interfere with cognitive processing. As you can start to come to terms with it, the arousal of those emotions becomes less powerful, which means that your cognitive part of the brain can start thinking clearly and that's when you get a greater sense of control over what's going on.

The most perfect situation is to be in a calm and alert state, which is what the parasympathetic nervous system can do for us. So, what we actually need to do is activate the parasympathetic nervous system to calm us down, so that the sympathetic nervous system goes quiet. Then we can do the things we need to do. In grief situations, there'll be a lot of things to think about. Too much. And, as time goes on, there will be a lot to organise, and it will be exhausting.

The big question is, how do we go about doing this? Well, as Dr Shaw says, there is no simple fix, but there are some basic things we can do to help kick-start this calming pathway:

If you don't express emotion, if you don't exorcise the grief, it could lead to further illness. It's so important to give yourself thinking time, time to actually feel what you want to feel, to actually think about what you want to think about. Just give yourself time. Go for a walk ... There's a great balance to be had between honouring the loss, thinking about the loss, giving yourself thinking time and allowing the parasympathetic nervous system to calm everything down, to get some exercise, to become busy doing something else that's productive that might give you a sense of self-worth and a feeling that you've actually achieved something – all of those things, small things on a daily basis make a real difference.

So it is that we can begin to adopt some practical approaches to grief. What I found most interesting from my discussion with Dr Shaw was how, by understanding what we're feeling throughout our lives as grief, we can adopt an entirely different response to our experiences of change and loss. Armed with this knowledge, we can now examine our lifelong relationship with grief.

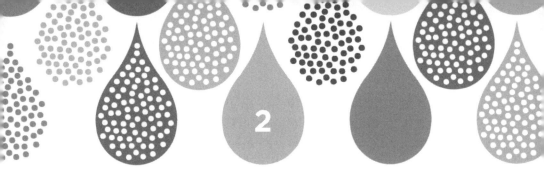

# GROWING-UP GRIEFS

The first time we experience a sense of grief or loss is when we embark on our very first journey into life – the exit from the warm, cosy sanctuary of the safe womb into this very bright and noisy planet we call Earth.

This transition we call birth is our first experience of loss and significant change. We have lost the safe cocoon that we called home for several months and now we find ourselves exposed, vulnerable, cold, hungry and very much out of our comfort zones.

Luckily, we are all born with a brilliantly simple and undeniably hard-to-ignore survival mechanism. It's a rather sophisticated alarm system to alert another person if our needs are not being met.

It's called CRYING.

As soon as we are born, we use our lungs to take a deep breath in and then we cry. That's what we all do as soon as we enter this planet: we announce our arrival into the world immediately, without any shame, without any holds barred – with all our might – by crying. Loudly.

We then keep going: we cry out when we are hungry, thirsty, too hot, too cold, uncomfortable, in pain, feeling exposed, overstimulated, tired, bored or need attention.

For new parents or carers, this alarm system can initially be very daunting, and to the untrained ear it seems to be permanently stuck on loud mode.

It can be a little bit confusing to establish what is needed to turn this 'alarming' sound off. Unfortunately, the baby isn't born with a volume control or a mute button! It usually requires a very patient process of mindful elimination with a trial-and-error approach to be able to figure out what to do and how to turn the alarm bell off.

We are all born with this wonderful ability to find homeostasis – to look for balance, harmony and comfort. It's comforting to know that we all have these innate survival instincts that are flexibly adaptive to the reality we happen to find ourselves in.

I was born very early – at thirty weeks, so ten weeks earlier than expected. I weighed just over 2 pounds and was very premature. As soon as I was born, I was whisked away into an incubator.

Of course, I don't remember this, but this was the story told to me as I grew up.

Because my parents used to tell me this story as a child, I must have embellished it in my immature brain with a mixture of made-up reality and odd dream-like metaphor.

I used to have this very vivid recurring dream as a child, and it was quite bizarre but strangely comforting.

I was alone, in the dark, bobbing around, maybe even spinning – it was hard to tell.

I was in a space with no obvious boundaries. I may have been floating, or sinking – there was no point of reference.

It was pitch black. It almost felt as if I had turned my vision around to stare inside my own brain and witness my own thoughts flickering on and off in the distance. Almost like when you have your eyes shut tightly and eventually pulsating balls of colour and lights seem to permeate the filmy eyelid that usually blocks the outside world from the retinas' scrutiny.

Was I dead? There was a heavy silence, a silence that was so irritatingly loud it was almost deafening.

It felt like I had no body, almost like being in a flotation tank that is exactly the same temperature as your body.

Suspended, in space, in mid-air? Where was I? I couldn't shout; I had no voice.

There was no one to hear me, even if I could. How did I get here? How do I get out? What should I do? What could I do? I could wait.

I surrendered to the stillness. A rhythmic reverberation punctured the stillness. What was that? I listened and felt.

It was a beat. A heartbeat. I waited, until it was time. Time for what? All these questions! Who was I asking them to? Myself? Who was answering them? Me? Was I having a dialogue with myself?

I realised, yes, I was. I was thinking, and this was all I was. I was a thought, a pulsating thought. A dream-like series of thoughts with a beating heart.

This recurring dream of mine felt very similar to grieving. It felt like I was grieving in my dreams as a child. And what I must have been grieving was the fact that I had no human-to-human contact for the first three months of my life. I was an incubator baby, and fed by tubes. It must have been a very alien feeling for a newborn baby to be transported from the warm womb straight into a see-through box.

I had this dream on a regular basis and always woke up feeling a lingering loneliness.

This particular memory recalled and triggered more memories, as if I were watching a cinema projection in my mind of my own life. It felt very surreal, and not dissimilar to the reports one hears of near-death experiences, where people talk about their life 'flashing before them'.

It was as if I too was scanning through my life's timeline. Searching for a point of reference, a nugget of familiarity, a moment of comfort.

After waking up from that recurring dream, I always wanted to be held, cuddled and supported. I wanted my mummy and daddy to be right there with me. It was as though I wanted to be back in my mother's womb, all warm, protected and safe.

I needed that feeling of comfort and security. When I woke up, I always felt like my life was very slippery, very fluid. I felt like I was melting, always melting back into the foetal position to find some comfort.

I always knew when I had that dream because I used to wake up curled up very tightly in a ball. Apparently, I slept like that a lot as a baby. On my tummy curled up with my bottom in the air. In yoga, they call this Bālāsana – the child's pose.

Even as an incubator baby I had developed a preference for turning over to sleep on my tummy. All my feeding tubes would get in a tangle causing all sorts of alarm bells to go off!

My mum always told me that I was the baby that caused the nurses the most grief in the hospital because I would only be happy sleeping on my front, and every time I was turned on my back, I would wail and cause chaos on the prenatal ward – so I think in the end they just let me sleep with my intubated face buried into the mattress. I was clearly a yogi baby!

All babies go through a period of separation anxiety. This can vary in intensity and in time depending on how it gets dealt with.

Separation anxiety is basically grief. An instinctive longing to be secure and feel contained, the fear of being alone is a survival mechanism – all human beings are born with the protective instinct for safety in numbers. The baby is programmed to cry and attract attention when it feels insecure and endangered. The baby has not learnt to look after itself yet and it doesn't have the forecasting ability to feel safe when it is on its own.

The trust and reassurance has to be built up over time. The baby eventually learns that the parent or carer has not actually abandoned them, that they are only a call away, and eventually the baby learns to be by itself for longer and longer periods of time. The fear of being alone is a feeling that a lot of babies and children have to go through to be able to gain independence and learn to self-soothe.

I was the first one among my peer group and cousins to have a baby (my son Krishan), so didn't have anyone immediately around me to ask for guidance. Both my mum and mother-in-law admitted that child-rearing was very different in their day, and that they didn't feel equipped enough to give me advice, which perhaps, in hindsight, was a good thing. So, I resorted to the internet, the baby channel on Sky and parenting magazines. I also read lots of child psychology books.

I must admit, after a while I felt extremely confused as there didn't seem to be consistency in the advice. There was so much conflicting information. There were theories that said one thing, and then studies that said the opposite – there was certainly no right way and lots of ways seemed to be the wrong way. I learned very quickly that the wealth of information I was consuming was useful up to a point, but the best and most effective way to spend my time was really very simple: to focus on getting to 'know' Krishan.

As I was 'hanging out' with him a lot, I got to learn his vocal cues. While listening and consciously paying attention to his cries, they became his unique language – the way he communicated his needs and requirements.

To an untrained ear he was just crying, but when I listened closely – like I had learned to – there would be different inflections in the cries, and the differences in tone and intensity were very subtle but they were there nevertheless.

There was an interesting variation in his crying. I learnt the subtle difference between the 'I'm very thirsty' cry and the 'I'm too hot' cry and the 'I'm really very bored now' cry.

He used to have this hilarious way of going to sleep: he would chant this rhythmical cry – it was more of a bleat than a cry. It would go on hypnotically for about fifteen minutes, because he was learning to put himself to sleep. This happened very quickly. He taught himself to sleep all the way through at 3 months old.

It was Christmas Day 2001 and I opened my eyes at 8 a.m. and panicked. I hadn't got up all night to feed him. Had I overslept? Was he OK? I ran into his room, and he was still fast asleep – on his back – his little belly rising and falling in a slow rhythm and his little chubby arms above his head. Totally fast asleep. Bingo. The best Christmas present ever: a baby who sleeps through the whole night.

He was such a brilliant baby: he had very advanced cries to communicate effectively exactly what he needed and when.

The 'nee-naw' fire-engine cry was his way of telling me he had done a number two, the slow ambulance wail meant he was in pain or discomfort and the rapid police siren was 'Feed me NOW!'

I realised that there was a very real language that I was learning fast. It was fascinating and, once we learnt each other's verbal as well as non-verbal cues, we started to communicate much more effectively and the parent–child trust became established relatively quickly.

Babies start to learn through emotion. They begin by reading our faces. They can somehow tell that when we are smiling this means 'happy' and when we are frowning it means 'sad'. They closely watch our expressions for cues and clues to navigate their new territory, and a friendly face conveys safety and an angry face conveys threat.

Once this communication channel has been effectively established between the parents/carers and children, it's almost as if the grief of not being understood quickly disappears and things start to get noticeably easier.

If you ask any mindful mother, they will all tell you that somehow instinctively they know what their baby needs – and this is because we, as mothers, have this instinctual ability to learn the language of our offspring. It's a very interesting dynamic. The mother and the child connection is incredibly sophisticated. As soon as we are born, we are learning. We are regulating, we are protecting, we are defending. We are doing everything we can possibly do to survive, and as human beings we develop some amazing methods and skills to be able not only to survive, but also to thrive.

Things really did change when I had my second baby. Jacob was born quite soon after Krishan turned 1. (I know, that was silly, but at least I got my extreme parenting done and dusted quite quickly!)

Jacob had a very different communication style. His crying seemed to convey some sense of urgency. The mistake I made was thinking that his language should be the same as his older brother's – so I didn't understand Jacob as easily as I understood Krishan.

Also, having a newborn and a 15-month-old wobbly toddler dented my patience and knocked some tolerance out of me, and my presence of mind was not as sharp. I was basically knackered, and the toll of two pregnancies in two years and having two babies under two had an effect on my mental health.

I began grieving my independence. Suddenly, in my late twenties, I had found myself a mother of these two boys, and I had no idea what I was doing and whether I was any good at it. I started to grieve my freedom.

Having a baby isn't normally linked to bereavement, because there is no death involved, only new life. However, I felt that there was a death of my old self. Somehow, I had lost a part of my identity and to me that felt very painful.

I craved my old life so much, especially as I was the first one of my peer group to have babies – I knew that all my friends were still going out and partying. They were all out drinking in the bars, they were all getting together and celebrating birthdays, and they were all having fun and being social.

I had just turned into this dumpy, frumpy person with no life, no excitement and who was full of resentment. Then, I felt guilty for being resentful, because I loved my babies. They were funny, they were delightful. They had an energy that was infectious, and they made me laugh. I was torn between who I was and who I had become. I was grieving my old self and growing into my new self.

Becoming a mother was not a natural thing for me. I hadn't planned to start a family so soon. I had just begun a new job – one of my dreams. I had bravely left the BBC because I realised that I couldn't move up fast enough in that institution, so instead, I decided to jump ship and to get a better job at an independent TV company.

I literally went from a coordination and admin role to a facility management role, and my learning curve was steep. But I loved that. I oversaw a lovely team of people and I even had my own office.

I had made it in my world. I was in charge of winning new business and was regularly found in fancy restaurants and bars, wining and dining prospective clients. It was the most perfect job for me – an impressive start to my career for my early twenties!

One morning, having been in the role for about three months, I got into work and sat at my desk. I felt unusually tired. My runner Laura came in with my cup of coffee and plate of jammy toast as she did every morning. She popped it down under my nose, and I caught a whiff of the strawberry jam.

She turned to leave and I instinctively grabbed her hand. She stopped and looked at me. I whispered, 'I really need Marmite.'

She looked puzzled. 'Did you just ask for Marmite?' she questioned.

'Yes please, Marmite.' I was very insistent.

'But, you don't like Marmite,' she protested.

'I know. Bring the whole jar please.'

As she crossed over the threshold of my office, I shouted, '... and a spoon!'

She scurried off, and shortly returned with the industrial jar of Marmite and a teaspoon.

I literally opened the jar and began eating the Marmite straight from it. As I was doing this, I looked at the label. In very clear letters on the front, it stated: 'ESSENTIAL SOURCE OF FOLIC ACID'.

*Isn't that what you need when you are pregnant?* I thought to myself.

I literally dropped the spoon and the jar of Marmite, grabbed my car keys and ran past a very confused-looking Laura, to whom I yelled that I needed to pop down the road and I'd be back soon.

Luckily, down the road there was a big pharmacy. I bought myself a pregnancy test, and in the work toilets, on that beautiful spring morning, I discovered I was having a baby. I was three months pregnant, and I literally had had no idea.

The growing up I felt I had to do in that moment was huge. I felt like my whole world had been turned upside down. From one new job to another new job in a matter of a few months. The adjustment period for me from discovering I was pregnant to Krishan being born was six months, and that was the steepest learning curve I have ever felt.

I had to let go of the old me and begin to get used to the new me. In that letting-go process, I grieved big time.

We aren't programmed to interpret these emotions as grief, because we are so hard-wired to associate that word only with death, rather than the concept of new lives, whether literal new birth or the letting go of a version of ourselves to allow a new one to emerge. Yet it is in these growing-up griefs that we begin to feel for the first time how profound an effect change can have on us and how societally we are so unprepared for the impacts of that on our emotional wellbeing. Many of these growing-up griefs can be centred around a feeling of being alone – a feeling that we are not taught to be at ease with, even though it can be hugely beneficial for us. Indeed, a 2017 study from the University of California, Santa Cruz on 'How to Be Alone: An Investigation of Solitude Skills' discusses the skills and positive impacts solitude can have.

## ONLYNESS AND LONELINESS

No, this is not a typo. It's a word that I made up for my imagination when I was younger.

Being an only child, my 'onlyness' was my best friend, my confidante and my companion. It was also my older brother and my twin sister. It was my portal into the past, the doorway into my future and the secret entrance into my very own wonderland.

I was Alice, I was Cinderella and I was the Princess who lived in the tower. I was a magical being who lived inside a treehouse creating spells and I was the fortune teller who read people's minds. I travelled to lands that don't exist in this dimension, and I could swim with the mermaids and fly with the angels. I was limitless.

I found that my own mind could come up with so many varied and outstanding stories, I would sometimes not know the difference between imagination and reality. My alternative realities were so incredibly colourful and entertaining that real reality seemed to pale in comparison.

My mother called me a compulsive liar, but I wasn't a liar, I just really *didn't know* the difference between what was real and what was make-believe. To me, the fairies did exist at the bottom of Diddima's garden, and I did go on many journeys with the White Rabbit through the looking glass and down the rabbit holes. There was a warmer version of Narnia in my wardrobe (I didn't like being cold), and the very tall tree outside did have a whole world of creatures living inside it.

Apparently once I told one of my primary school teachers that I was a secret princess and I had been adopted by my current parents because my real parents, who were royals in a faraway land, had died. (That may well have been a combination of *The Little Princess* and *Aladdin* – who knows ...)

All these things were true for me, and I was not lying: I was just very, very good at reading stories and becoming part of them. Then all the stories I read merged inside my own head and I was actually living in them. I imagined it all so deeply, it became my reality.

I really do believe that if this was today's world, I would have been diagnosed with some kind of psychosis or a type of ADHD. But, back then, compulsive liar seemed to be as good as it got.

Adults didn't understand me. My teachers would constantly complain to my parents that I used to tell too many fibs at school, and this unfortunately became something that everyone started to believe. Dipti was the girl who cried wolf.

There must have been many more examples where my imagination ran wild, and I don't remember them all – unfortunately, my parents are not around for me to ask – I only have little snippets of information, vague memories and stories they told about me to others, so who knows if any of this is true.

Our memories are not accurate: they are essentially little stories filed away in a drawer, and when they get dusted off, they either get eroded or embellished – some words and letters go missing, because time has erased them, and we do our very best to fill in the gaps. Sometimes other things have grown on them, kind of like a beautiful mould. In this process, some inevitable editing occurs and then the memories become altered. Recalling memories for me is like playing a solitary game of Chinese whispers that just happens within my own thoughts.

Indeed, back then my tales were exactly like Chinese whispers; the messages and stories just got slightly distorted and beautifully embellished along the way. I didn't 'tell tales' in the way they were accusing me of 'telling tales'. To the adults around me, telling tales meant I was spreading untruths, which I guess technically they were, but in my mind I was not doing anything bad because my tales were full of delight, wonder and magic. The tales I told were incredibly creative and transformational.

I was, and still am, a deeply honest person, and honesty is one of my core values. I didn't like being called a liar and a fibber. I have a vague memory of being around 11 years old and someone important – it could have been a doctor – telling me that I would have to take some sort of medicine to help me stop telling lies. I was a very bright child, and I think it was at this point – when I was faced with taking 'truth pills' – that I made the incredibly difficult decision to simply turn my vividly colourful imagination off. Aged 11, I just shut it all down.

This made me sad because the far-fetched tales had become my world, my own way of escaping. My imagination was happy and never let me down or abandoned me. And now I had to watch it go dark, wither away out of my head and disappear out of my life like my goldfish Frankie. No words, no explanation, no goodbye, just gone. Just like that. Flushed down the toilet.

That's when something incredibly important inside me died, and I really don't think that I have ever stopped grieving it. It's that child-like part of all of us – the spark of creativity without boundaries that we are all born with.

I think that every single one of us will grieve our freedom to imagine. All of us, at some point, will either unknowingly or be forced to suppress that sparkle of child-like light that we are all born with. Inevitably, as we get older, it gets dampened down or blown out by well-meaning adults, especially when we hear those dreaded words: 'you just need to grow up now'.

My onlyness was an incredibly gifted friend, and my trusted companion for just over a decade of my life – it was my inner world and, without it, my imagination was a dark and empty place. So, when my imagination was blown out, my onlyness very gradually turned into loneliness.

When I hear the word loneliness now, I remember my own feeling of otherness but, with my awareness of how powerful and helpful onlyness was as a way of reframing being alone, I am able to help people who see loneliness as a problem to find their own way of turning it into their version of onlyness.

Ultimately, from the moment we're born to those later transitions as we start to identify what kind of person we are, we experience grief in those losses of identity or changes in narrative that come with growing up. The comfort of our prenatal womb-like state is gone and we feel cast adrift, but it is often that reframing of the narrative from loneliness to onlyness that can help us see grief as a natural part of the growing-up process, and help us to understand that the grief becomes a powerful part of our story.

According to Harvard's Center on the Developing Child, during the first few years of a baby's life their brain will form more than 1 million new neural connections every second. It is now also widely acknowledged that the first five years of a child's life are crucial to their brain development, and that as we move from infancy through toddlerhood into our early pre-teen years, our brains are still evolving and have a high level of plasticity, which means that they are shaped by our learning and experiences. If we keep these facts in mind, then we can safely assume that if things are changed or lost during this key time we may develop some beliefs and subconscious behaviours in response to the stories we are told about ourselves, our environment and our lives. Therefore, it is crucial that we begin to establish healthy ways of discussing our emotions around growing up, being alone and loneliness, loss and grief during these very early years, to enable us to cope with the change and loss that comes as we move through life.

# EVERYTHING CHANGES

When examined in a safe space, grief can reveal so much about who we are, and why we behave and act in the way we do. While on the surface it can appear to be a destructive force, when faced with it – up close and personal – it is also a mechanism for survival. Grief can sometimes be the problem and, ironically, grief can also contain the solution – as a powerful tool to connect us human beings back together again.

If we learn how to look at grief with an honest filter, we can gain some incredible insights that can help us no matter how old we are. Change and loss affect us all through our lives. However, the most pivotal time is through our childhoods into our teenage years, during which our bodies and our understanding of the world around us undergo profound changes. From the first five years of exponential brain growth, as we move through into adolescence, we experience a period of brain 'refinement'. According to the UK National Institute of Mental Health, the brain is at its largest as we enter the teenage years and thereafter it begins to mature, with certain neural connections being 'pruned' away and others being strengthened. The teenage brain is also incredibly plastic, which means that it's receptive to changes in the surrounding environment, and these can directly impact the brain's development at this crucial stage.

This chapter focuses on a fictional mother and daughter – let's call them Jaya and Charlotte – with two very compelling stories. Divided by grief initially, they are then miraculously reunited through the very same grief that separated them. Their story is upsetting and, although fictionalised, the struggles that are narrated here are very real for very many people. At the back of the book (p. 222) I offer some helplines and support networks that might be useful if you – or anyone you know – are experiencing something like this. It is also vital that if you are encountering issues of this nature, you seek professional mental health support and advice.

## JAYA'S STORY

'No one really understands, no one really listens. No one gets me at all, in fact none of them know me. I'm not like all the others. I'm paid to smile and pretend like everything is not twisted inside my head. My life isn't my Instagram feed. The filters remove the bloodshot eyes ... Thank god for those filters ...'

She's saying these words as her phone vibrates constantly in her back pocket. She pulls on the cuffs of her oversized grey hoodie and picks violently on the loose threads of her distressed boyfriend jeans. There's an interesting scraping sound coming from her feet as she occasionally rubs the zig-zagged edges of her Adidas Yeezys together. I continue to listen. There's a long pause.

Then she looks up at me and laughs nervously.

She is a 21-year-old Instagram influencer. She has just hit 1 million followers on her account, but instead of feeling elated she feels despairing.

Like most of my clients her age, she has been sent to me by a well-meaning parent. Jaya's mum, Charlotte, got in touch with me because Jaya started having severe panic attacks.

In many of the cases of severe anxiety and panic attacks that I've dealt with, often the person is emotionally surprised by the panic attack because it strikes at a moment when ostensibly they aren't in distress. For example, frequently people report to me having a panic attack in the middle of the night or sometimes at moments of rest. Interestingly, when they say this they rarely use the term 'panic attack'; in fact, most likely they've felt as though there was something physically wrong with them. They might describe experiencing a fast-beating heart, feeling dizzy, nauseated or light-headed and having sweaty palms. Of course, these symptoms can indicate a physical problem and are always worth getting checked out by a medical professional. But they can also indicate a deep anxiety and a sense of loss of control that has been unacknowledged and is now manifesting itself in a very scary and real way. It's emotional overload – very similar to a fuse blowing. When this happens, it's not uncommon for people to end up going to A&E with a panic attack, and that's nothing to be ashamed of.

In Jaya's case, panic had struck out of nowhere when she was in her bedroom at home. She was scrolling through her online accounts when her heart started racing, her palms became so wet she could barely hold her phone and she thought she was either going to be sick or faint. She called her mum for help. Charlotte ran upstairs and had no idea what was happening nor what to do. She suggested Jaya put her head between her legs and do some deep breathing. But now Jaya was doubled over and worrying about whether she could breathe deeply or not. She started to hyperventilate. She was now in full-on panic. This was not Charlotte's fault – she didn't know what Jaya was going through and was trying to help her as best she could. However, when someone is experiencing a panic attack, they are emotionally overwhelmed and anxious, so telling them to breathe deeper can exacerbate this. Instead, it's a good idea to use these five steps:

1  Walk them to a private place (if they're not there already), stand in front of them
2  Look into their eyes and ask them to watch your face
3  Talk to them, facing them with clear and kind instructions
4  Hold them firmly by the shoulders, reassure them and repeat 'You are safe'
5  When they are looking at you, ask them to repeat with you – all the time holding and facing them – 'I am safe'

Eventually, shaken and exhausted, Jaya's panic attack passed. Generally, panic attacks last anywhere between five and twenty minutes, but it can feel like far longer when you're experiencing them. The episode made Jaya and Charlotte realise that they needed help to understand what was going on.

Most of the time, people in their twenties do not want to be in 'therapy' or told what to do. They want to be *seen* – and when I say seen, I mean *really seen*: understood, heard and held.

They want to be loved. Being loved by a million followers or fans may seem like the perfect recipe for affection, but it is not. We all need to be loved by real people who know and understand us, and ultimately by the one we see looking back at us in the mirror.

Hypnotherapy is never usually the first port of call for anxiety or depression – I am usually the last stop. The doctor is usually (and rightly so) the first place to go; the doctor may then recommend cognitive behavioural therapy (CBT), talking therapy or medication.

Jaya had been to the doctors and been prescribed diazepam, which she did not want to take. She had sessions with a CBT counsellor and, although it had helped, it just hadn't completely 'clicked' for her.

Sometimes thoughts create more thoughts, which in turn generate stories and feelings. Feelings help us regulate our emotions and link to activity and action. We make a choice about the feeling – is this safe, dangerous or neutral? Is this bad or good? Do we need to take action, sit still or hide? The feeling then drives a behaviour. The behaviour creates an experience, and finally our experiences shape our realities. Therefore, if we want to change some aspect of our reality, we must be willing to examine the thought that preceded it. This level of self-enquiry is what I begin to teach in my practice.

A person suffering with extreme anxiety is most certainly not able to stand back and look at their life from a high level of self-enquiry. They are completely stuck in the grips of their own primitive mind. Also known as the limbic system, this is the original, primitive, emotional part of the brain, and around this has developed the neocortex – our evolved and intellectual mind. The brain is a very curious place.

If you understand how it works, and you know the role you play in managing your own mind, you can begin to take charge over your whole life.

The primitive part of the brain is responsible for keeping us safe, secure and alive. It is the part that is protective, emotional and defensive. It is very much problem focused and in charge of risk assessment. It is responsible for our levels of motivation, emotion, learning and memory, as well as desire and pleasure. It moves us forwards or holds us back.

Over time, above our primitive brain has formed what we call the neocortex and the prefrontal cortex. The neocortex is the part of the mammalian brain involved in higher-order brain functions such as sensory perception, cognition, motor commands, spatial reasoning and language. The prefrontal cortex is located at the front of the frontal lobe. It helps us with complex behaviours, including planning and positively forecasting, and greatly contributes to personality development.

The brain is a complicated machine with a complex structure, and the paradox is that it has very simple functionality in terms of human behaviour. If you understand how the brain operates at this level, you can start to change behaviours and thought processes.

The limbic system's primary function is to keep us safe, protected and to warn us of danger and threat. When it senses danger, it triggers a reaction in us that we can't ignore – anxiety, fear or anger. All these emotions are designed to be strong and loud to attract our attention so we can get to safety as quickly as possible. If we were primitive people, living in caves, we would need our primitive minds switched on to high-alert mode. If a tiger wandered into the cave, we would have to react instantly to preserve our lives.

We would quickly choose between fight, flight or freeze. I, personally, would not choose to fight a tiger, or attempt to run away from it, but freezing may be a good option. Playing dead and staying as still and quiet as I can might just save my life. If the tiger senses I am no threat and I'm not its plaything, it may lose interest in me and wander back out of the cave again. Freeze, in our world, can take the form of self-doubt, denial, disbelief, procrastination and depression.

Let's just pull the covers higher over our heads and hope the world out there disappears. The world out there is dangerous, full of threat and difficulty, and this means I need to stay safe – deep inside my cave, under my rock or, in the modern-day context, under my covers.

In the same way, a suicidal person is very much in extreme flight mode. The suicidal mind is feeling completely overwhelmed, confused and numb. I see it as grief: the person is grieving the life that they want to have and feel like they can't achieve. The person is already grieving their own existence as they contemplate ending their life.

Their brain is overreacting to life and their constant questioning of their life. Their positive belief systems have been hacked. The suicidal mind is like a virus – it's aggressive and attacks the 'healthy' mind from within. The suicidal mindset is convincing, strong and completely powerful. A person in this state can no longer see a way out and doesn't know what else to do.

They don't know how to ask for the right help, or if they have attempted to ask for help in the past, they will feel that they were misunderstood, unheard or perhaps let down in some way. They feel lonely and alone, and often the irony is that their suicidal mind is actively pushing everything and everyone away so they can be alone. They think of themselves as unworthy and un-loveable, and have either never learnt how to fully accept and love themselves, or they have forgotten how to.

Their self-esteem is at rock bottom as they convince themselves that them not existing will be better off for everyone involved. They have lost the ability to think about the consequences of their actions or to think clearly about how to solve their issue. They are completely driven by emotional overwhelm and intense grief.

All they will hear in their mind is a loud voice telling them that they need to get out of this situation, this experience, this life – there is no other way out, this is the answer. They will not be in tune with solutions; they are very much focused on their problems.

All my clients, when they start coming to me, fit into the spectrum of fight, flight or freeze, or a combination of all three. They feel anxious, scared or angry, and this will make them want to retreat, be defensive or inconsolable.

It's my job to understand and fully listen. It's my job to get to know the person in front of me by paying attention to their likes and dislikes. It's my job to help them untwist their thinking. It's my job to help them vent, safely.

Jaya explained to me that the more followers she got, the more anxious she felt. She talked about imposter syndrome, feeling immense pressure to show up and perform even if she hadn't slept for weeks. She didn't eat properly, she always felt weak and tired. She had been suffering from eating disorders that started just after her eleventh birthday – as puberty hit. Now, her body dysmorphia had spiralled out of control to the point that she couldn't look in the mirror at her real face any more. Her Instagram face was now her mask and she hid behind it.

It took her just under ninety minutes every day to put her 'face' on. Her real face hidden underneath the layers, the contouring, the sculpting and the artistry. As she talked to me candidly, barefaced, I asked her how she was able to come and see me with no makeup on.

She looked confused and sat very quietly thinking. I could almost hear her brain searching for an answer and then she softly replied, 'I feel safe here.' Finally, a place where she felt safe enough to be free. A place where her skin could breathe freely too.

Jaya and I developed an amazing therapeutic relationship very quickly. She said I was the first therapist that didn't make her turn her phone off, insist that we drag and dig up her past or make her cry.

Solution-focused therapy works in a clever but subtle way. We gain information by asking solution-focused questions, and we don't dwell on history; instead, we understand the history as a way of learning how certain beliefs and models of the world have been installed.

Like the search history on a computer, this can give us a picture of the content that has been shown to that person, and this content is what shapes our views of the world and what we believe. These beliefs get digested and formed together as our truths.

In the several months that we worked together, Jaya came to her own realisation that her need to be followed, liked or accepted came from her childhood, when she was encouraged to perform and be on show.

As a toddler, she had a perfect cherub-like face, but she was very shy – she didn't start speaking until she was almost 5 years old, and then words were few and far between. She was just a quiet kid.

Charlotte, her mum, wanted to help her with her extreme shyness and develop her confidence. So, she signed Jaya up to a child modelling agency, and she very quickly became one of their 'popular' children.

Charlotte left her job as a freelance PA to become Jaya's PA. Jaya spent most of her childhood travelling around the country and sometimes the world, posing for photographs, getting dressed up, sitting around in studios, beaches, villas, hotel rooms, being gifted all sorts of products and freebies. She was always on show – with her mum by her side. It looked like an amazing life, for both of them.

Jaya was very good at following instructions – she did that very well. The modelling agency didn't care about her shyness behind the lens: they said that if she could stand, sit and smile on demand, that's all that mattered.

This way of life was all Jaya had ever known. What a lot of people didn't see was the tears, the pleading and the anxiety behind the scenes. Jaya began bedwetting around 9 years old, and secretly sucked her thumb. These signs of childhood regression are symptomatic of feeling unheard and unrepresented. The body is always showing up what the voice is unable speak.

Children need to feel secure and there must be rules and boundaries put clearly in place for them to know where they stand and why that boundary is there. If the boundaries are very loose, or very tight, they feel a desperate loss of control as well as a feeling of being over- or under-controlled. When I see children in my practice, it's quite often because they have been given too much of a voice or not enough of one.

When a child is given too much of a voice, they can become expectant, greedy and entitled. They run rings around those caring for them and use their voice in a controlling way. The tantrums become their weapon, and if tantrums are not dealt with in a consistent way, the child very quickly learns that further tantrums help them get them their way, and the boundary between carer and child gets blurred.

When a child is unheard, they either feel like their voice is not relevant, and they can become reclusive, or it can swing the other way, as they want to make sure they are heard: they can become loquacious or even rude in their desire to grab attention.

Jaya negotiated reduced hours with her modelling while she was doing her GCSEs, and she said that that was her happiest time. This was unusual, as a lot of teenagers come to see me because of exam pressure and academic stress. She told me that that was when she was left alone. She enjoyed learning and using her brain to think for herself. This was the only time she could express herself through her writing and she felt free.

She did well in her GCSEs and, in the summer holidays before she started her A-levels, she was offered her first mega Instagram influencing job.

It was paid very well and her mother felt that she couldn't turn it down. She was earning a few thousand pounds a day, and this was certainly (on paper) an amazing opportunity. Jaya shakes her head as she talks about it.

'When I see these people going on reality TV shows, desperate to become an influencer or a public figure and all that they are after is that little blue tick of verification. I just know that they are all suffering silently, like I am. I look like I have the perfect life – easy fame and instant publicity handed to me on a plate – that should make me happy, shouldn't it? Mum constantly tells me how she has dedicated her life to help me be where I am today. She is always saying I need to be more grateful and that I don't know how lucky I am.'

Over our time together, Jaya learned to turn her attention away from blame and victimhood, and to find her own inner compass, pointing her in the direction of authenticity, trust and kindness.

She told me things that she had never revealed to anyone before and, week by week, she felt the metaphorical weight shifting from her mind as she began to put some healthy weight back on. As the weeks went by, I could see remarkable changes in her language, her appearance and her personality.

Her periods returned, her taste came back and her self-esteem grew. She could turn her phone to silent and became comfortable with silence. She also developed a new voice. A voice that she felt like she was hearing for the first time. It was a quiet voice, but very strong. She understood that her quiet strength was growing from the positive seeds that were being planted in her awareness. The seeds that were now being sown in her mind were the metaphorical flowers and we had begun the process of pulling out the weeds.

Jaya started taking control over her posts and started to share more of her real self online. She began to do her own thing and began raising awareness for body positivity and self-love. She did lose a significant number of followers as her message began to change. However, this seemed to be a relief to her rather than a worry. She described it as cathartic.

The most incredible thing was that her relationship with her mum had also started to noticeably improve as they were able to sit down and have heart-to-heart conversations with each other about how they were both feeling. They were starting to share experiences with each other and explain things to each other in a non-confrontational way.

Charlotte began to open up to Jaya in ways that she had never before, and this was a powerful healing process for both of them to understand and connect with each other.

Jaya was so happy that her mum had started to see her side of things and understood how she felt. This was a huge relief to her, and Jaya felt that her sessions with me were done.

Charlotte wanted to speak to me about something once I had finished seeing Jaya, so just after Jaya's last session, we arranged to have a chat.

# CHARLOTTE'S STORY

It was a really lovely moment when, after Jaya had stopped coming to see me, Charlotte called and thanked me for everything I had done for her daughter, but also told me about how I had helped her deal with some of her own trauma.

Charlotte explained that she had been grieving and in a very bad place for two decades, and the grief and negativity she was experiencing had been twisting her thoughts – and, through her daughter's recovery, she noticed the personal transformation in herself.

She talked to me about her own life and revealed that her boyfriend had left her when she was seven months pregnant with Jaya. He left in the night, packed a small bag with his essentials and was gone.

It was like he had died, but she knew he wasn't dead. His possessions were still around, and a month before she gave birth, she had no idea if he was coming back, or how she would manage financially as well as emotionally.

Her family didn't support her: they blamed her for getting pregnant so soon after meeting him. They were Catholic and were not able to accept this baby, born out of wedlock, as their grandchild. Charlotte and the unborn child were cast out of their lives. Charlotte suffered intensely with low self-esteem and post-natal depression.

She had to raise Jaya on her own, with no support, no income after she gave birth and feelings of complete hopelessness. She told me about a time when her landlord served her an eviction notice because she hadn't been able to pay her rent, and she contemplated suicide. The only thing that stopped her was her 1-year-old baby.

Charlotte was trying to be two parents at once and, through her sheer desperation to make money any way she could, she was humiliated, and was forever grieving her own dignity.

She realised that she got obsessed by Jaya's success – she wanted to show Jaya's father, her former boyfriend, what he was missing out on with the success she was creating through their daughter.

Jaya never mentioned her father to me – she told me she never met him. I had no idea what the story was there. It all made sense hearing Charlotte's point of view, and understanding her motivation and drivers of behaviour helped Jaya gain a huge insight to her mother's own hurts and grief.

When Jaya plucked up the courage to courageously admit to her followers how anxious she felt, the reaction of many was cold. Some called her an 'attention seeker'. She lost thousands of followers from that post. The brand that she represented decided that she was no longer the best 'fit' for their clothing, and her contract was not renewed.

Jaya was not seeking attention: she was grieving her privacy. Charlotte was not being a pushy parent: she was simply protecting her own dignity and self-respect that she had grieved the loss of for so long.

After hearing both their stories, it was clear they wanted the same things.

Every human being needs these simple **LAWS** in place:

To feel **L**oved
To feel **A**ccepted
To feel **W**anted
To feel **S**afe

If these fundamental laws are threatened, we feel grief.

If love fades, we immediately miss it and there is a deep sense of heartfelt loss until we can connect with it again.

If we are not accepted for who we are, we feel wronged and grieve our sense of self.

If we are not wanted, we feel rejected and grieve our sense of belonging.

If our safety is compromised, suddenly we feel exposed and vulnerable and grieve our security.

Grief, in its multifaceted way, is our incredibly sensitive alert system to help us realise when we have come out of homeostasis – essentially it is trying to be helpful, to return us back to a feeling of belonging, benevolence and balance.

I have a question for you ...

While you were reading Jaya's story, and before you heard Charlotte's story, what was your opinion of Charlotte?

Did your opinion of the situation change after you heard Charlotte's story?

We, as human beings, instinctively make choices and decisions and form our opinions about what is bad, good, negative, positive, agreeable or distasteful as soon as we hear a story.

Our belief system and emotional mind get caught up in that story and begin to wrap themselves around the storyline that is presented to us.

We don't naturally step outside of that story to imagine another perspective unless we make a conscious effort to or are supplied with it.

This is called personal bias. We all like to think of ourselves as open-minded and would not like to hear that we are all biased in some way. However, the brain has developed our personal or unconscious bias system to keep us protected. Our biases are formed based on our life experiences, our learning and the way in which we have been treated. Our biases are designed, essentially, to keep us safe.

I would imagine that, as you were reading Jaya's story, you may have been slightly critical of her mother. Perhaps you formed a negative opinion of her without realising why. But, then, as you began to read Charlotte's story, you perhaps saw and understood the whole situation with more compassion and a wider perspective.

Similarly, when we are offered opposing versions of life, side by side, this can be a useful way to overcome our natural personal biases and allow us to develop our own sense of perspective.

This idea of being neutral, impartial and unbiased comes in useful when dealing with emotional overwhelm such as grief. Grief, as a system, is naturally and overwhelmingly emotional. The dramatic emotional pull of grief is designed to get our attention. It's the brain's way of assembling some kind of order and structure to keep us on high alert or hidden away or fleeing. Back to that classic fight, flight or freeze.

During storytelling, the brain flicks out of intellectual control and logic and gets caught up in the story the emotional mind is telling us. The emotional mind either helps us to connect, unite, bond and collaborate with each other, or it forces us apart, disconnects, removes and severs us from each other. It does this with the underlying primitive reason of safety. This is why stories have always played such an important part in forming communities from the earliest days of humankind.

♦♦♦

Growing up in the last century – a teenager in the 1980s – I felt like I was living in two confusing systems: I physically lived in working-class West London, while I was also expected to inhabit my parents' Indian middle-class metaphorical culture and thinking.

My parents had got married and had me within a few years of first meeting. When I was 5 years old, they settled down and made Harlech Gardens – a nice enough council estate in Middlesex – their new home. This is where my memories really begin to feel real, and I started to make connections with lots and lots of people.

Harlech Gardens was spread out, with 240 flats divided up neatly into little pockets of about thirty flats. Each block had eight flats in it, which were populated with mostly down-to-earth, working-class people who called a spade a spade. We did have some very interesting neighbours over the years though.

I think the crazy variety of people I was surrounded with while growing up led to my fascination for and love of people watching.

There were certainly a lot of curious people to watch, and as I had my very vivid imagination for company, I used to make up some hilarious stories using my neighbours as walk-on parts.

I had so many characters to choose from, it was like a director's dream, but it must have been my parents' worst nightmare. I can see that now they are gone. But, while growing up, I used to be very confused as to why I was never allowed to hang around with any of these people – they were so interesting to me.

I don't remember there being much trouble on the estate. It was a nice place to grow up. I had so many potential friends on my doorstep as there were loads of us estate kids and we had so many different places to hang out – the disused garages, the warehouses at the back, the sheds and the parade of shops opposite. Around us there were loads of other council houses – different estates, with so many more kids potentially to make friends with – but I was very rarely allowed off our estate, unless it was to ride my beloved Melody (a blue bicycle that I got for my tenth birthday).

I tried to disappear as much as I could – I loved visiting and used to sit for hours listening to my neighbours, who would tell me stories about their lives. The Polish ladies were like my estate grandmothers, and their eyes used to light up when they heard Melody's bell ringing outside their window.

I really liked visiting the very friendly Spanish family at the far end of the estate, who always drank their coffee in sweet little transparent jewelled glasses, until one of their sons tried to be too friendly and show me 'something special'. Then, there was the group of crafty ladies who organised the jumble sales and they were also the ones who knocked on doors recruiting volunteers for the summer fêtes and the jubilee celebrations, for which I always volunteered.

There was the lovely family from the Philippines who seemed to have a permanent BBQ going on their balcony; the ones who walked their dogs at night, because no one was really allowed to have pets; the ones who would always keep themselves to themselves and never opened their curtains; and then there were the curtaintwitchers – and there were plenty of them. Nanny Phoebe seemed to the oldest one on the estate; she didn't have any family and loved having me over to visit – she called me her 'darling angel'.

Nanny Phoebe used to leave me treats in the planter outside her door on a Wednesday because that was the day her shopping was delivered. One Wednesday, there were no treats. Then the next week, none again. The third week, as I looked in the planter that now had a wilting plant, I decided to knock. No answer. When Phoebe died, no one told me. I used to go there, and nobody would answer. I had a feeling that she had died, but when I enquired, it seemed to be a very touchy subject.

My parents didn't know her, so there was no point asking them, and also, if I did, it would have been a dead giveaway that I went to visit people when I wasn't allowed to, so I knew better than to stitch myself up! So, I asked my other neighbourly friends – particularly the ones who may have known who she was ... Sadly, all of my neighbours were suspiciously reluctant to talk about Phoebe. They would either avoid my question, pretend they hadn't heard me, or change the subject altogether. Nobody seemed to be comfortable talking about death or about Phoebe. I found that very odd. It was like she hadn't existed.

Then there was Mrs Benson, who lived two flats above us, and I only knew her because she used to 'walk' her big black dog very early in the morning, then sit on the small wall outside my bedroom window, smoking. I would wake up to the incredibly loud sound of her relentless coughing combined with the nasty stale smell of her cigarette smoke wafting into my room, and have to go and shut my window before my parents thought I had a secret morning smoking habit.

One morning, there was no coughing. No coughing for an entire week. I knew Mrs Benson had died. It was obvious to me. I told my parents that I thought she had died, and they thought this was a very strange thing for me to comment on, given I didn't 'know' her. It turned out that for the whole week her dog had been constantly barking inside her flat, and eventually the neighbour below her called the council. She had died, alone. Lung cancer. My parents never spoke about her again, and nor did anyone else. It was like dead people seemed to be deleted from alive people's memories, and I didn't understand that at all. It was something about the culture in which we lived, but I was too young to know whether it was because of class or because of where we came from.

My parents, despite being surrounded by working-class people, were not working class; they were middle class but somehow, due to their circumstances and health, could not afford a British middle-class lifestyle, and this difference in our family was startlingly obvious. We were the 'poor' ones and, quite often, I was sent to stay with my cousins in the suburbs during the holidays so I wouldn't have to socialise with the estate kids.

Both my parents had come from middle-class backgrounds in their native lands.

Growing up, my mother lived in Kenya, her father was a doctor and they had a good life. She talked about being well cared for, she mentioned her housekeeper and she remembered being a happy child. Until she was 8. That was when her father died of cancer and, being the only girl, she was sent away to a boarding school in Shimla, India. She had only vague memories of this time, but she did tell me that she felt very lonely.

My father, one of thirteen, left India in his thirties with a very good engineering degree; he was ambitious and adventurous. He told the story of leaving the blistering 40-degree heat of Calcutta and arriving a few days later in Germany where it was -40°C that year. He only had on a thin outer jacket and was carrying very few possessions. In the ten minutes or so that he stood on the station platform in Stuttgart, where he was to meet his friend, he literally froze, so much so that he had to be defrosted before he could bend his arms and legs enough to sit in the car.

Beyond these snapshots, my parents didn't talk to me about their past. I remember asking lots of questions because I wanted to turn their past into a historical novel inside my mind and use their experiences as part of my tales about my past, but they seemed to be reluctant to speak about any of it. Even when we had neighbours around for drinks. (My dad would keep a bottle of Mateus rosé wine in the fridge, just in case a neighbour popped in.) The neighbours would ask questions because they were curious about this Indian family who socialised with them and did everything they could to try to fit in. We were told, as a family, that we weren't like the other Indian families on the estate, who stuck together and kept to themselves. We were 'different'. I liked hearing that. I liked being different. I was a black sheep and a Bounty bar. In fact, 'Bounty bar' was what I got called a lot – brown on the outside and white inside. Being a Bounty bar (or a coconut) was the only way I could fit in and, because we as a family didn't want to be different, we all became coconuts. My dad refused to let my mum teach me Hindi or Panjabi. He said that I should only speak English, because otherwise I 'would get confused'. When he said this, I used to wonder what he meant because so many of my friends were successfully bilingual and managed not to be 'confused'.

I think he was trying to protect me from racial discrimination, and he often encouraged us all just to blend in. But I didn't want to blend in. I liked being a black sheep-coconut. I liked standing out.

There was a difference between standing out and feeling out of place, and I used to notice this subtlety when I went to visit my relatives outside of our estate. There, I did feel out of place, in the nice green leafy suburbs of Pinner and Woodford Green. No one else seemed to be out and about whenever I went to visit my cousins. I couldn't understand why no one knew or hung out with their neighbours. The nice and neat cul-de-sacs were always manicured perfectly and there were all these individual houses all nicely spaced out with immaculately kept front gardens and huge back gardens that were privately fenced off. None of my wider family had to share their lawns – that was a weird concept for me. The lawn outside our flat was a shared space, and our 'block' always hung out there together. It was like a massive outdoor party every summer holiday. I felt very at home in the council estate, but I was brought up to believe I didn't really belong there. So there was always this element of confusion. I wanted to fit in – but I was never allowed to.

The community school nearest our estate was, in my dad's eyes, a less academic school, so he insisted I went to the school 5 miles away. It had a better reputation, but this also meant that any new friends I made did not live near me and, again, my friends became inaccessible. So, as ever, I was encouraged to just be on my own.

When it comes to growing up, we must quickly work out which tribe we belong in, which community we feel safe in and which relationships give us a feeling of familiarity.

Starting school, particularly secondary school, can be a very tricky time – a time when we are starting to question who we are and where we fit in. Some children can be out of their comfort zone and it's remarkably easy to get lost and ultimately lose direction.

I was the first one into my tutor room on the first day of school. My dad taught me never to be late. So, I never was.

Mrs Thomas (not her real name) was my tutor. She was a very cheerful and bubbly Welsh lady. She had a strong accent and I loved the way she said my name – rolling the 'i' at the end and elongating it for a little longer than necessary.

'Diptiiiiii,' she repeated, when she asked my name. 'What a delightful name! Since you are the first one here, would you like to be Monitor?'

I agreed, not really knowing what that meant, but I felt special being asked so I said yes. It turned out that this involved collecting the register every morning and afternoon to bring to the tutor room. That was it really, but that made me feel useful, so off I went, as instructed, to go and get the register from the office, which was in a building on the other side of the school, before the rest of the class arrived.

On my return, I got lost trying to find my way back to my tutor room, which was also the needlework room, and found myself wandering around on the second floor, where the science labs were, way after the bell had rung. Eventually I worked out that home economics was on the first floor and made my way down there. The corridor was consumed by a pungent scent of burnt casseroles. On one end of the corridor there was a large cooking room and on the other end was the needlework room. I remember being worried that I had been gone ages, and I could feel my heart beating in my feet.

Finally, I found my tutor room again, where I was supposed to be, and I slowly opened the door. The room was now busy and buzzing. I was so anxious, and this is the reason why, thirty-five years later, I distinctly remember walking into my tutor room for the second time on my first day in secondary school like it was just the other day.

The outer tables on the edge of the classroom were sewing machine tables, so no one was sat at those, apart from one red-headed girl, sat on her own in the corner.

All the central tables were now full. As I stood at the doorway, I noticed an interesting pattern. There were seven Indian girls plus Caroline (not her real name), the tallest girl in the class – they were all sat across the first two tables, which comprised a long table with stools joined to a smaller square table nearest the doorway. There were the confident, chatty four together on the small table, and the three timid ones and Caroline sitting on the long table with the stools. All five popular-looking white girls were sat around one big table in the back right corner.

The boys were divided up on to two tables. There was a smaller table of four Indian and Asian boys, plus Dean (not his real name), in the middle of the class who looked like they loved school, and the other larger table at the front opposite the door housed five white boys who all looked like they hated school.

Then there was Annie (not her real name), in the back corner.

Mrs Thomas announced my arrival in her Welsh jolliness: 'Aaahhhh, there you are Diptiiiii, did you get lost already?'

Everyone looked at me, and I wanted to run back out. But I didn't; instead, I nodded, handed her the register and she said in her sing song way, 'Sit wherever you want to ...'

I went over and sat with Annie in the back corner, who looked marginally irritated, I thought.

How incredibly tribal we were at 11 years old. I found it fascinating that, even at that age, most of the class (whether consciously or subconsciously) chose to sit in groups that matched their ethnicity or skin colour. This is likely down to our tribal programming – this idea that we find the people that feel safe and, because they resemble us, they have the same vibe as us and perhaps match our way of thinking.

This seating plan was not formulated by the school – everyone had made a subconscious choice of where they wanted to sit without necessarily thinking about it too deeply.

If I hadn't been the last into the classroom, I would have still chosen to sit on the edge over in the far corner. That felt safe over there, because you could see the whole classroom, and it was at the back of the class by the big windows and was the sunniest spot. Perhaps it was also because I subconsciously, somewhere along the line, had already decided that I belonged on the edge and, as an only child, I am quite comfortable being on my own.

I found secondary school a very difficult transition. Looking back, I realise I did not fit in. There was a part of me that was so jealous of the popular Indian girls, who were always making private jokes in Panjabi and had their own 'in thing' going on.

Even though it was the 1980s, I never felt that there was any racism in our tutor group; however, some of the Indian girls were quite racist towards the white girls, but they didn't see it at racism, they saw it as sticking together. They had no desire to mix with the white girls, even though they always looked like they were having fun. I desperately wanted to hang around with the popular white girls, but the Indian girls looked so disapproving of them, I didn't want to upset anyone.

I couldn't hang out with the boys, although I looked like one. I had stupid hair. It was so short and spiky that when I wore trousers every single teacher mistook me for a boy. Before I started secondary school, I had beautiful wavy shoulder-length, layered hair, with a lovely thick fringe – I looked rather cute. I think my parents were worried about me looking too cute, so my mum insisted I get it cut like a boy, and told the hairdresser to do exactly that.

So, I probably could have got away with hanging out with the boys, but that just was not the 'done' thing at 11 years old in the 1980s, and very tall Caroline's table just didn't feel quite right – although that was probably the one I did belong on. They were the misfits. However, I wasn't ready to accept that I was also a misfit. I really wanted to be part of a tribe, but I didn't fit into any of them.

So, it was just me and Annie, but Annie was most definitely a loner, and found my chattiness a 'bit too much', so quite often I was by myself. I didn't belong anywhere – I felt drawn to the loner, because I didn't want to be alone. I was technically a nomad – a nomadic coconut (brown on the outside, white on the inside).

From my first day in secondary school, I felt like I was in deep grief. I was grieving for a tribe to belong in, and I was grieving my lovely fringe and my wavy, layered hair. It wasn't until we started studying for our GCSEs that our tutor groups began to get mixed up with all the other tutor groups. That was when I found a tribe to hang around with but, by then, I had already decided that school was an unsafe place.

Our primitive mind gets switched on to keep us safe. The feeling of safety translates into a sense of belonging. A need to fit in, or in some cases stand out and be seen. We all have this unconscious tribal need to be part of a gang or peer group. Then, within those groups, very quickly do we find that a structure of social and interpersonal hier-archy naturally develops. Often this breaks down into a leader, the enablers and the informants, and the remaining are loyal followers.

As we grow up and we leave the tribe and hierarchy of the family environment, an interesting thing happens to us. On one level, we are free to choose a new tribe, but we are equally free to choose to be on the fringe.

Teenagers are always learning who they are by getting feedback from their friends, their parents, their caretakers and their teach-ers. They are naturally rebellious, because they are breaking out of a mould of protection. Think of it like a butterfly emerging out of a chrysalis. A little delicate creature, now with these brand-new wings, begins to recognise that they have freedom, and that freedom for some can be hugely liberating, and for others, it can be utterly over-whelming, depending on the personality and how deeply and how strongly they have been conditioned to think in their early years.

This state of emergence can trigger a deep grief reaction in some, which they may not realise or recognise as grief but is actually bound up with lamenting their lost childhood identity and grieving the changes that have been thrust on them physically and emotionally by adolescence and the fast pace of change that comes with moving through the 'goals' of the educational and social system. This is heightened by quite literally what is going on in their heads. Interestingly, the American Academy of Child and Adolescent Psychiatry talks in their Facts for Families series about how the latest brain imaging reveals the frontal cortex in an adolescent brain is underdeveloped and in fact doesn't fully mature until we're about 25.

This frontal cortex 'deficiency' matters because it is this part of the brain that controls more 'rational' thought and decision-making, which means that the adolescent brain is more reliant on its emotion-driven amygdala (the part of the brain that is employed in our fear responses and more 'primitive' behaviours). It is this dominance of the amygdala over the still-forming frontal cortex that means teenage responses tend to be impulsive and emotional. In fact, in his book *Emotional Intelligence*, Daniel Goleman refers to this as 'amygdala hijack'. This sense of the brain and body being hijacked by our emotions is often central to our experience of grief, and explains why the issues we experience during this teenage stage of life can often be bound up with grieving for not only our lost childhood identities but also for the identities that we want to carve out for ourselves as we emerge into adulthood, but often can't due to the expectations of others.

The Treaty Centre opened its life-saving doors when I was 15. It is the main shopping centre in the middle of Hounslow High Street, and within it they had built a huge and very modern library. The Treaty Centre Library had a huge sectioned-off study area with about twenty individual kiosk-like desks where all the dedicated students could reserve a kiosk and sit and concentrate without being disturbed. Unknown to those who planned it, this library proved to be the key that we teens needed to unlock the metaphorical cage formed by our parents' watchful eyes.

The library obviously had books and resources in it (remember, this was before the internet), so we all did have to go to the library to get any information we would need. But the brilliant thing was that back then the library was open until 6 p.m. every night, and on Thursday it was late-night shopping, so it would stay open till 8 p.m.

This meant that when school finished at 3.20 p.m., I would go to the 'library' until it shut and get home by 7 p.m. This would give me a few more hours of liberty, and then, of course, on Saturday the library was open and that's where I would always be – without fail. As far as my parents were concerned, I almost literally lived in that library. I sound like a right studious swot, don't I? Well, if I let you into a little secret – promise you won't judge me?

Off I went, with my folders full of paperwork, my pencil case, my books and my cardigan (the air conditioning in the library was always set to much cooler than it should have been). I would find an empty kiosk and place my stuff in it. Then I'd open my file next to a pile of books, with one book opened up and placed carefully spine up to denote work in progress. I'd usually scatter a few highlighters and pens about, pop the cardigan over the back of the chair, and then immediately leave.

Funnily enough about 75 per cent of the kiosks were occupied by paperwork and similar faux work set-ups but were completely empty of people!

There were about fifteen of us 'library goers' who discovered this freedom pass system, and some of us would hang out together, wander around the shops; sometimes we would sit in the café and sometimes we would literally just walk up and down the centre checking out the 'talent'. Sometimes, some of us would disappear for the whole day because we were having secretive rendezvous ... it was brilliant. We were known as 'The Dossers' – I'd found my tribe at last.

Dossing in Hounslow High Street for most of my teenage years was my way of rebelling and gaining independence from my parents, who I never felt understood me.

The system of pretending to study and be a complete swot was working so well until, one day, the librarians got really fed up with us using the library as a free bag drop and cloakroom facility and started confiscating our stuff. Signs got put up all around the study area with a sixty-minute warning. All unattended belongings in a kiosk after one hour would be confiscated and there would be a £5 fine to retrieve them. Basically, our books were being towed away! To get around this – because I didn't want to stop dossing, but equally, I didn't want to hang around with all my heavy school bags and folders – I had an idea. I noticed that there was this other girl who would always be there when I was and, like me, arrive, dump her stuff and disappear, so I approached her and suggested we share a kiosk to halve the fine. (I had no idea that I was an entrepreneur back then!) She loved the idea, so the two of us would arrange to meet at the same time and squeeze both our stuff into one kiosk – and then we would only have to pay £2.50 each to retrieve our stuff.

Sonia (not her real name) and I shared a kiosk for a few years, and we became good friends. Her parents were traditionally Gujarati. They were deeply religious and, even though they had two daughters, they had converted the largest second bedroom of the already small maisonette into a temple, and the sisters had to bunk in with each other.

Sonia was going out with Steven (not his real name). Steven was 19, and he had a car and a job and splashed the cash around quite a bit. They seemed like a nice couple and were clearly besotted with each other.

When I asked Sonia how she met Steven, she was always a bit reluctant to carry the conversation on, until she finally admitted that, weirdly, Steven had gone out with her older sister, Nina (not her real name), at school. She waited for my reaction and was relieved that my response was totally non-judgemental, so she felt she could confide in me more and more.

She explained that the reason Nina and Steven had broken up was because Nina felt that she couldn't tell her parents about their relationship, because she felt she had to listen to her parents' wishes about her getting married to someone else that they had picked out. Being the older daughter (she was three years older than Sonia), she was sure that her parents wouldn't approve of Steven because, firstly, he was not Gujarati and, secondly, he was white.

Steven used to sneak in to see Nina in the sisters' shared bedroom after their parents had gone to sleep, and that's how he got to know Sonia. Sonia had always had a huge crush on him and knew that her sister would have to give him up one day, so she made her feelings known. Sonia couldn't tell Nina that she was with Steven, because she knew that Nina didn't want to get married to the other guy.

Her parents – even though they had chosen to move to England and bring up their girls in West London – still felt strongly that how they approached relationships in their own culture 'back home' was right and could not accept any difference to this way of life, as they regarded any deviation as a massive threat to their belief system.

Even though I had never met Nina, Sonia invited me to one of her wedding events. Asian weddings are elaborate – there are about three or four events that precede the wedding, then there is the wedding itself and then a huge celebration and banquet afterwards. To a first-generation Asian parent, the size and magnitude of the wedding is a direct reflection of status. In fact, for some, Asian weddings are more about the status symbol and less about the people getting married; they are so grand and, the more people there, the better it appears. So, all sorts of people end up getting invited to all the events, really to make up the numbers: neighbours, shopkeepers, friends of friends, people you might have met at the bus stop once or twice – anybody who will come – and this helps with keeping up the appearances to the rest of the community.

Then what happens is all these people, who don't know the bride and groom, go not *for* the bride and groom, but for a good old nosh-up and free drinks. As this was a Gujarati wedding, there was no booze, but they made up for it with copious amounts of free food!

I used to be so confused about this craziness and how the hell these normal working-class people used to make this financially crippling extravaganza possible. Usually, the families would be very frugal-looking, often with small houses, battered cars, no luxury items. So they would save, save and save – saving every single penny to throw this enormous wedding to enhance their family name. Literally all their savings would be spent over a few days, and it would be over. Twenty years of life savings completely blown in one week.

When Sonia introduced me to Nina, we were at her Mehendi party. She looked stunningly beautiful. She was sitting on a jasmine-covered podium, adorned in jewels, wearing a gorgeous crimson lehenga dripping with gold; she was in the middle of getting her hands and feet tattooed with an intricate henna design. Nina looked up and smiled at me with her perfectly made-up lipstick mouth, but behind her big, beautiful, kohl-lined eyes, she looked totally dead.

Later on, that evening, just before the dancing was about to begin, I walked into one of the hotel bathrooms, and I heard soft sobbing coming from inside of one of the cubicles. I stood very still and waited for a few seconds. There was a bit more sobbing and then a hysterical-sounding voice. I don't know who she was talking to but what she said has stuck in my mind for all these years: 'This isn't my f***ing wedding. This is my f***ing funeral.'

I realised it was Nina, and I didn't want her to know I had over-heard that, so I quickly walked out and stood outside door thinking, *Oh my God, what a thing to be saying, the day before your wedding day.*

To what extent our past shapes our future depends on how much we let it.

Looking back and grieving for a past that wasn't as we hoped it would be impacts our present and affects our future. This way of thinking is problem- rather than solution-focused, and it keeps us stuck.

In this case, and in many others I know of – and there have been many – people have been pre-programmed and pre-conditioned from a young age to believe that their lives have to be a certain way. If they don't develop an ability to stick up for themselves – because of fear of rejection or being ostracised – they end up living a life that is not *really* theirs.

Then grieving the life unlived – the one that they *really* wanted – becomes embedded into their everyday existence and feeds into unhappy relationships, erratic behaviours and poor mental health, without realising that what underpins all of this is unacknowledged and unaddressed grief.

For others, being knocked off a predetermined path comes with a different sense of loss, one that has been felt by so many young people this year due to the global pandemic. The lives of schoolchildren and university students have been derailed in ways that none of us could have imagined, and that has meant that so many of the long-anticipated rites of passage have been taken away. I caught up with my son, Krishan, and his friends, Liv and Ezra, to see what impact not doing their A-levels has had on them. All three of them are now at one of the UK's leading universities:

### Ezra

I was just working really hard throughout year 13. And I had coursework for history and geography. And that's the majority of all the work I'd been doing up until the pandemic came and A-levels were cancelled. And when A-levels got cancelled, immediately I was really, really happy, as I'm sure most teenagers were.

From my mocks, I had my predicted grades, and I was fairly confident that I'd be able to get my place here [at university]. And that's all that mattered to me. So, I wasn't really bothered about getting a billion A*s, I was just bothered about getting in. And I was fairly confident I would do so. So, I was really happy. But I know a lot of people who weren't in that position. And they weren't happy. So, there was some disappointment even for me because, I guess, for two reasons. One reason is I sort of thought that if we hadn't done A-levels, we might be discredited as a year group and I was kind of sad about that. Because even though you could get three stars, and I got my grades, and they were really good. Would everyone think 'Did I lie on my CV?' and I'll probably still be looked upon as the year that never did the real A-level study. And there's nothing I can do about that ... So, for example, I've got the same A-level grades as my older sister, and my oldest sister is here as well [at the same university]. And I sort of still have this underlying feeling that, even though we've got the same grades, people – my parents, for example – it doesn't seem like they think I've actually done it, although they've never said that or even implied it. It's probably just me thinking that. But there's a certain sense that I never actually did it.

I was also disappointed especially because of my geography coursework, I put so much effort into that I spent my whole summer doing it. It was 100-page document, it was basically a dissertation on the cohesion between ethnic groups in my local area and whether that affected political voting patterns, and the quality of the environment in my local area. And I got nothing back for it; it wasn't even marked.

I know a lot of people who basically predicated the fact that they would pass their exams on this coursework. So, one of my friends basically did absolutely no work throughout the whole of year 12 and year 13. And then really worked hard for this coursework, because they were like, 'Damn, I really need to pass my A-levels. I'm going to work hard for this coursework.' And they obviously got nothing back for it, and it was probably the only credible piece of work they ever did in their A-levels.

### Liv

So, I didn't take an entrance exam to get here. Then I didn't take any A-levels. And I don't really feel like my GCSEs really reflect anything. So it feels like I literally have done nothing to get here. I have a very big family. I saw some of them over summer after my grades. And I was the first one to get into [this university]. But they were like, 'You didn't really did you?' They actually said that and it was quite raw. And also my grades, they were like, 'Well you didn't really get them.' It's because we're quite a competitive family.

I don't necessarily feel like I've been cheated [out of the experience of doing the exams]. I feel like I'm cheating. It's like I'm waiting for them to say 'Oh, what are you doing here?' Because I didn't, I haven't, done anything to get in. I feel guilty.

*Krishan*

I had the same thing with my CV, when writing down [my A-level] grades, I didn't feel like they were my actual grades. But then for my GCSEs, it does a bit more. So, it feels like we never actually did them.

There are other rites of passage that come with the move from leaving home and going to university that have been eroded by the pandemic, and in our chat we touched on whether these new university students felt that they had been 'robbed' of the opportunities to go out, find new social groups, establish their own lives in a freer way.

*Ezra*

Yes. Simple answer is just yes. Especially in these last few weeks. I didn't feel it so much in the first few weeks, especially because pubs were open, but the last two weeks especially.

[Generally, though,] I'm just glad for the fact that nobody in my family has contracted Coronavirus badly. That's the main thing. I'm grateful and thankful for that.

I think the thing which has helped university life specifically, is that we've come here having no idea how it was before. So we've been thrown into this place. And this is the norm for us, whereas I think it's probably harder for the second and third years. For example, speaking to my sister, because my sister knows what she's missing out on. I have no idea what I'm missing out on because I've never had it. And we know we'll probably have it at some point. But for my sister, she's a third year now. She's almost missed out on basically half her degree. Because she had to do her last term in lockdown at home and she's had this term disrupted by lockdowns. And the next few terms, she's probably not going to be able to socialise as much as she usually can. So, I think she's probably been affected a lot worse by this than I have.

*Krishan*

Freshers [the university welcome week] was fun because everyone's still getting to know each other, and we made things work. But the last few weeks since the lockdown started, it's become quite boring.

But I do think we had an accelerated getting to know each other process because everyone felt really united by a common enemy of COVID.

### Liv

I had one day of Freshers, and then had to go into isolation. Obviously, then you're dealing with the fear of missing out, especially because everyone was meeting everyone. And I was worried that everyone would have made their friends and that was it. While I'm just sat in my room and watching it on social media.

I've made up for it now. I don't think it's actually really impacted my friendships or anything. But in those two weeks, I definitely thought everything was happening without me.

Also, to follow on from what Krishan said about how we just got to know each other really quickly, compared to other years maybe. Equally, I think that's drained a lot of people's social batteries quicker than it might have done in previous years, because you feel like if you're in your room, you're not socialising and you're putting yourself in isolation. When normally I don't think people would have thought that much about whether to go out or not tonight, but now it feels like you're missing out on something really big.

Talking to Ezra, Krishan and Liv, it was surprising how resilient they were. Even though elements of the guilt and uncertainty that they touched upon are very much part of grief's suitcase of emotions, they didn't identify as grieving themselves. However, there has been much written about the huge mental toll that the pandemic has placed upon the student population across the age ranges, and, although only time will reveal what has really gone on in terms of impacting their personal development, it is hard to imagine that there will not be an outpouring of grief at some point for these lost fundamental experiences of growing up. We'll come back to looking at the pandemic specifically in the Global Grief chapter, but it's important to remember that, as we move through these age-defining parts of our lives, we come into conflict all the time with what we thought life was going to be like when we got to be 16, 18, 21, etc., and what it is really like, and that, as much as we grieve the reality of things lost, we are also grieving the vision of things we never had too.

As the pandemic stretches into another year of our lives, Jacob, my other son, has also had to let the idea of doing his A-levels in the normal way go. As I write this, he is currently studying for exams that will be marked by his teachers and the whole examination structure that they have been used to has now broken down. I asked him if he feels like he is grieving and he replied that he didn't know any better as he has never done A-levels like they are supposed to be done. He feels the grief and the sense of social isolation because he and all his friends have not been able to see each other and socialise outside of school. His year group have been essentially learning all of the curriculum fully online on Microsoft Teams and Zoom calls. The majority of his A-level schooling and teaching has been through online classes and home-working. It's been an unprecedented experience for the classes of both 2020 and 2021. What a strange final year of school. Jacob said that he knew that this would be part of the history books, and that made him feel marginally better.

♦♦♦

For many of us, there comes a point when we start to figure out how to let go of all the stuff that our parents or society wanted us to be, or the rituals that we thought defined who we were. Then we can carve out our own pathways by holding on to the unique stuff that makes us who we are and we recognise that we can look for the signpost signalling the right pathway for us. Once we figure out how to walk our own path with our head held high, and learn to stand up for ourselves, it's as if we're somehow liberated. This is because we have worked through that grief and see that the life unlived was simply part of the journey to the life we were supposed to be living all along.

# FOR BETTER
# OR FOR WORSE

Human beings naturally want autonomy. We enjoy the feeling of having control over what happens to us in our world. This reassuring idea of having authority over our lives, our thoughts and our behaviours keeps us feeling safe and secure, and gives us a defined structure to work within. The concept of identifiable and conscious control gives us parameters that we can use as templates and boundaries to keep us flexibly contained, supported and held. Indeed, many of us grow up believing that we will reach this sense of autonomy over our own lives when we become adults.

When my clients come to see me, it's typically because they're experiencing one of two feelings, or both simultaneously:

1  They feel that they have fallen out of the system or container they were once in and find themselves in a place of limbo and uncertainty.
2  Or, conversely, they feel that the structure or system they are part of has locked them in, leaving them trapped and unable to find the way out.

This idea of loss of control and change from familiar routines can have a huge impact on their maps of reality; if they find themselves in new territory with no obvious plan in place, their mental faculties, their abilities and their capability to navigate through their lives easily and effortlessly can feel seriously inhibited, and then their reality can become punctuated with fear and anxiety. Essentially, they are in deep grief for a framework of reality that has fragmented, either because something has directly impacted it or because it has turned out to be entirely different from their vision or understanding of it. In the same way that we've seen those emerging into adolescence often grieve for what they've lost in their childhood selves, adults often grieve the future vision they had of their lives that is lost to reality.

They talk about blocks in their system, feeling stuck or overwhelmed by a situation, or talk to me about repeating loops and behaviours that generate unwanted experiences. This type of language suggests they feel trapped, limited and at a crossroads, and have lost their way – they have been going around in circles, like on a roundabout with no signposts, and just can't see which exit to take. They feel disorientated and each time they go around the same circle, more confusion sets in and the feeling of 'I'm lost' kicks in and, eventually, they end up sitting in front of me.

With clients who have lost their way and feel directionless, I like to use the sat-nav (GPS) metaphor. I say to them that solution-focused hypnotherapy is simply a navigational aid that can help point them in a direction; they are the driver of their vehicle and must programme the destination that they want to head to into their GPS, and once they have done that, I can offer them a road map with a suggested route.

I remind them that they are in full control of the pace they go at and they will also be able to make the choice to take the scenic route, the fastest route or even the route that avoids any tolls, and this choice is flexible and can change as we go through the sessions.

The client is always in control, even if they think they are not. Solution-focused hypnotherapy simply helps them to remember that they are. They must make a commitment to the process by agreeing to enter into a contract with themselves that they are willing to collaborate and cooperate with their desire for change.

Human beings are very good at making contracts. Contracts are a way of giving us a system of codes that we can agree to and sign up for. Some contracts are verbal and honoured with trust and faith, and others are legally binding or flexible, while others are more serious – like an oath taken in a courtroom, a declaration of allegiance or even signing the Official Secrets Act. At all levels these contracts act as a secure and agreed system that helps us all know where we stand and allows us to feel settled.

I signed a book contract before I started writing this book. With that signature, I formed an agreement with the publishing company that I would deliver a certain amount of words to them by a certain date, and for this they would pay me a sum of money. This is a typical contractual transaction – an agreement to provide a service and an agreement to accept the terms and conditions set out in that agreement.

Would it surprise you to discover that once I had signed the contract with the publishers, even though I wanted to write this book and I knew what I was signing up for, I still felt an equal amount of happiness and loss of freedom?

The excitement that I initially felt was being mixed with a sense of sudden overwhelm and concern about my precious time that I would be spending and losing.

Notice the language there. I had formed a story inside my head that I was giving up time – which I view as a precious commodity. If we feel we are giving something up without our volition – whether that is a title, a position, a person, or our own time – we may feel a sense of loss of something valuable.

Interestingly, this triggered feelings of loss and grief within me. I immediately noticed this and, as soon as I stuck the metaphorical grief label on the feelings, they suddenly felt better. Since this book is all about grief, my feelings about committing to writing it are an excellent illustration that grief isn't necessarily felt in bad times or times of trouble, but that we can also feel grief in good times and when positive things happen too. This distinction is very important to bear in mind – grief as a feeling can universally apply to the good and the bad.

I was projecting a sense of loss for the time I would be 'giving up' in the future to devote to the writing of this book, and this feeling of loss generated a belief in my subconscious system that this agreement to write the book would have a negative impact on my autonomy.

This is a classic case of a 'negative thought virus'. Once a negative thought enters the mind, it can have a detrimental knock-on effect to all sorts of systems in our bodies, and this then creates a negative loop that can reinforce itself. The powerful thing to do in this situation is to notice the thought virus and catch it before it takes hold, infects and overwhelmingly spreads to the whole of your life.

This feeling of negative overwhelm is what I am labelling as grief. It's a sudden overload of negatively charged emotional questions that pop up out of nowhere and have no place to go. Like releasing a can of unpredictable worms, they wiggle their way into our awareness and we can quickly feel at a loss and spiral out of control – just like our reactions to a pandemic.

Little did I know that, a few months after signing the book contract, we would be going into full lockdown because of the coronavirus pandemic. As I write this today, we are still in full lockdown; life is totally different – our liberty and in many cases our incomes have been taken away – but the commodity that seems to be widely available to most people right now is ... time.

Time is a certainly viewed as the most important of all commodities when you are on your deathbed, yet most of us spend our lives waiting around, wasting time or using it in ways that we know don't truly fulfil us.

The pandemic and nature of how things have rapidly changed has certainly brought about a general feeling of loss of control and that has resulted in a lot of anger, frustration and other heightened emotions.

On the flip side, there are a lot of people who have been ever so grateful because of the time they have been gifted with. People have been able to take a massive pause from the hamster wheel of life. Many have been shifting in headspace and have seen how important their families, children, relationships and homes are.

There have been epic levels of creativity and ingenious ideas born from this pandemic period. There has been a massive increase in spiritual awareness. It's almost as though the whole planet had a huge wake-up call. Some people heard the alarm clock and got up, and others pressed the snooze button and fell back to sleep.

When our idea of personal control is taken away from us, we either resist it and push back with anger and cries of injustice, or we move into creativity and find a way to do something differently. The loss of control can have a very different outcome depending on your outlook on life.

I have learnt that grief is a system of feelings that is essentially unavoidable. We may know it and experience it differently with different nuances and names, but as soon as we can recognise it and identify with it, it feels familiar, and this level of familiarity is somehow strangely comforting. This familiarity also allows us to apply strategies and principles to help us break down the feeling and deal with it using our informed, educated and logical minds. This is where the adult brain has an advantage over the teenage brain in that, by this point in our lives, our prefrontal cortex is more developed (usually reaching maturity around 25 years of age), and this is the part of the brain that allows us to apply problem-solving to the whirl of emotions being processed in our amygdala.

Once I heard myself say I was 'giving up my time', it raised a red flag. I know that this language is problem-focused and, because of my solution-focused training, when a client tells me what they feel they would be giving up or losing, my immediate response is to ask them what they will be gaining or how they will be benefiting instead.

Similarly, I needed to change the language in my mind surrounding writing this book to locate the gains and the advantages. Once I could ask myself these solution-focused questions, I was more able to fully engage with the terms and conditions of the contract I had signed and shift my perspective to a positive frame of mind. That accomplished, I could wholeheartedly submerge myself into this project and tap into all my initial excitement and enthusiasm.

I told myself that my time can still be appropriately managed by me: I am fully in charge of how I divide my time up to devote to writing this book. I had to remind myself of the value I would receive and the value I would be giving by dedicating my time in this way.

Hopefully this gives you an idea of how, when the language we use tilts into a solution-focused frame, things feel lighter and become uplifting, and the concept we previously found oppressive begins to feel possible, do-able and exciting. Ultimately, motivation then becomes your rocket fuel, and that is a lovely feeling. It really is as simple as updating our thoughts, so that loss becomes gain, grief becomes celebration.

I updated my original thought of 'writing this book is going to take up too much of my precious time' to 'I've been given an opportunity to devote my time in a new way that is beneficial to myself and others'. This new thought has a much better feeling attached to it, and when you feel better about something, you are much more likely to be more adaptive and behave in a positive way. This behavioural flexibility is very powerful, and it is a skill that we can all develop when we know about it. Once we know how to update our thoughts in a positive way, things can positively change in our lives.

One of my mentors, James Tripp, author and developer of *Hypnosis Without Trance*, teaches a very simple but powerful system of how when a human being can update their thoughts, with the use of their imagination, they begin to feel differently, and this new feeling can alter their physiology and elicit a change in their sense of reality. This reality alteration is often a catalyst to identify possibility and curiosity, and ultimately leads to creating incremental and even monumental changes in their lives. He calls this system the Hypnotic Loop. James explains it in such an easy-to-follow way that makes total sense. I have been using a slightly adapted version of James' Hypnotic Loop formula and solution-focused questioning combined with taking my clients into deeper states of relaxation, which we can call trance, to create a solid foundation for any type of change work to occur safely and with meaning. This is my adaptation of James' Hypnotic Loop:

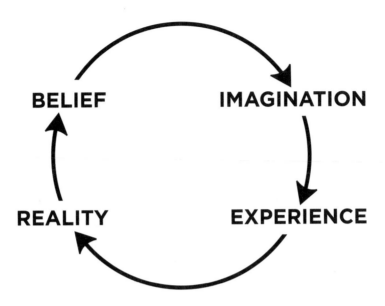

Often, by the time we reach full adulthood we have developed our own set of personalised belief systems which are a series of thoughts that we have adopted and adapted to be true. When we have a thought that we believe, we use our imaginations to make that thought come alive for us. Once the thought becomes conscious it most likely has an associated feeling attached to it and the feeling creates an emotion. This emotion helps us figure out how to behave – it can be a driver to move forward or a reason to hide. An indicator that something doesn't feel right or even a cue to change direction. This movement in our behaviour – backwards or forwards or even standing still – shapes our actionable realities: what we can do, can't do or find ourselves doing or not doing.

A powerful example of this brain association is a simple moment I had with my eldest son Krishan. Krishan used to have hay fever as a child. One day, I was showing him some photos of crop circles in Wiltshire, and as he was looking at the photos, his eyes started to get itchy and nose started to run, and he began to sneeze. We were both very surprised at his immediate hay-fever-like reaction to the pictures; he looked quite confused and said, 'I think I'm allergic to these photos.'

The brain is certainly a phenomenally interesting place and the imagination is so very powerful, which is why it's vital that we know how to get it working for us rather than keep allowing it to work against us. The brain is a very complicated network of intellectual and emotional communication pathways. However, even though the brain is complicated, there are simple patterns that drive certain behaviours.

For example, a feeling of thirst can result in a decision to go and get a drink. A feeling of being too hot can result in removing a layer of clothing. These simple feelings produce simple automatic behaviours that require almost no debate or analysis. It gets slightly more complicated when our emotions can be used against us in a negative way.

I am interested in exploring this idea of emotional hijacking because it is so important to create awareness of what can happen if we let our emotions get the better of us. My aim is not to frighten you, but to educate you so you can understand why it is so important to be in control of your emotions, rather than the other way around.

## HIJACKED BY GRIEF

James Bore is a cyber security expert, and he tells me that almost all cyber security attacks, also known as computer hacks, are possible because the hacker knows how to play around with people's emotions.

The computer hacker goes for a swift emotional hijack using fear, worry or guilt so that the person immediately changes their behaviour and gives away information – a password, personal details or even direct access to a bank account – so that the hacker can wipe the account clean or take control of the computer to get what they want. This type of hacking happens all the time on an individual level and is also happening on a massive scale to large companies and organisations.

However, there is a much more sinister type of hack which is a long-term hijack that uses a sophisticated strategy. It mimics the behaviour of a biological virus – entering the system and sitting dormant for a few weeks while spreading through contact and infecting as much as it can before the host exhibits any symptoms.

Similarly, a cyber virus enters the computer and knows how to remain undetected in the system, and while it is in the system it can be doing a variety of things. It can be spying, copying information, deleting information, changing data, spreading and infecting other users. The time that the virus enters the computer and remains undetected is called 'dwell time'.

On average, this type of stealth attack can take up to a year. In that time the hacker, who is quite often a person, has a plan for the attack. Think of this person like a computer terrorist. They want to cause a motivated and targeted disruption. Once the network has been penetrated by a hacker, they will quietly map it out. They will be trying to understand it. They will be trying to compromise the security of those computers by turning the control over to themselves as stealthily as possible. They will be hiding information and at all times preparing for when they are detected and will have a quick getaway planned.

They may also remove log files, which helps them remain undetected. While they are in the system, they need to be as quiet and as subtle as possible to avoid setting off any alarms.

As James was describing this to me, I had visions of a game I used to play as a child called Operation. The game consisted of a cardboard image of a body on top of a metal circuit board. The cardboard body overlay had these little gaps with the vital organs and bones sitting in them, like an Adam's apple, a rib and a heart. The idea of the game was to use the tweezers to remove the little parts and organs without sounding the buzzer.

This hacking 'operation' is big business. Once the dwell time is up, and the compromise has been detected, then it often gets more serious and can be hugely damaging because the attacker then has nothing to lose; that is when you hear of cases involving ransom-ware – where all the data is encrypted so the company cannot access any of their information, or passwords lock people out, or even sensitive data and information is made public. Essentially, information and data is kidnapped, and you need to pay to get it back.

James relays some tragic stories of people who have lost family photos of loved ones who have passed away. The hacker knows this and holds these precious photos to ransom for a huge fee, but even if it's paid, they are often not returned. Sadly, we are increasingly hearing of similar tactics being used in revenge-porn cases in the news, but these sorts of emotional ransom attacks are rarely reported. James says they may not be deliberately targeting grieving people, but they do know that grieving people are likely to be hugely emotional and therefore are 'easy prey'.

Have you heard of phishing? These are scam emails sent out to a vast list of email addresses aiming to get money. Like actual fishing, the hacker or 'phisher' is casting their net wide and loading it with emotional bait, in the hope that this will attract a good catch.

Such attacks are also carried out over the phone: that 'accident' you had, that claim you haven't received yet, that promise of a pay-out due to that huge injustice you faced – all connecting with your feeling of being a victim.

The vast majority of phishers use emotions such as fear and urgency. For example, you need to do this thing immediately and, if you don't, you will get in trouble, e.g. a warrant for your arrest is being issued by the Inland Revenue, your house will be repossessed, you will be put on a black list, or you will even be frozen out of your account.

Alternatively, they also use reward, opportunity and good news as a mechanism: you've won a prize, you've been chosen, you have got a pay-out ...

Some sophisticated ones seem so realistic, like only yesterday, I got a telephone call supposedly from an online company that I have an account with. The lady was very nice: she told me that that my yearly subscription was coming to an end and they had a good offer on. Instead of re-subscribing online, they could offer me a much better deal if I paid over the phone. Normally it is £85, but over the phone I will save £35 and it will only be £50 for the year. Wow, that's a great deal. Or is it?

Imagine if I didn't have my wits about me: I would have given her my card details with all the information, like the three numbers on the back, and my account would have been wiped clean. Instead, I told her that I was aware that this was a phishing scam and I would be reporting the number immediately. She hung up.

Can you imagine how many people she called that day? Say she called 2,000 people, and 10 per cent of them fell for it and they each paid £50. That adds up to £10,000. Not bad for a day's 'work'.

There's another type of phishing attack that uses shame as a trigger. This is called scareware. You may get an email saying that there has been a video taken of you in your most private moments, and if you don't pay X amount, they will send the video to all your email contacts and friend list. This can easily play on a person's consciousness and this scare tactic, using shame and guilt, works very well, and is highly lucrative. It is horrific emotional hijacking – short term and long term – as it feeds into a cycle of shame and fear.

I asked James how we can tell the difference between a real threat and a pretend threat, and he gave me this brilliant piece of advice: as a rule, if they do have something on you, they will send it immediately as part of the threat – to show you evidence. At that point, go to the police. Please do not fall into the trap of being manipulated.

There is a brilliant system that some government agencies use as part of their security clearances when employing a member of staff. They ask for full disclosure of any blackmail material that may be held against you – for example, if you were arrested for possession of marijuana as a teenager. This isn't for them to use or hold against you, it is so that if any other person was to hold it against you, using the threat of your position or job, the fact that you were fully transparent when taking on the role means any disclosure would not affect your position. Obviously, this depends on the seriousness of your 'crime' – but the point is that this transparency gives you a certain level of freedom in knowing your past can't affect your future.

The lesson here is about being brave and owning up to those skeletons. The upset, the shame or the secrets you may be harbouring are actually not that important (or interesting) to anyone else apart from yourself. Once you are okay with revealing yourself with no shame, secrecy or guilt, you can be freed up in so many powerful ways, and it gives the hackers much less control over you because you have claimed greater agency back.

What does this all have to do with grief? Well, the grieving process is an unavoidable brain hijack in many ways. The brain's virus protection is now compromised – physically, mentally and emotionally. We can get ill, we can have negative thoughts and also our emotions can be hijacked. To prevent all this doing long-term damage, we can put some practices in place to enforce damage limitation.

Like an attack on a computer network, once it becomes clear that the system has been compromised, there are processes in place to minimise the negative impact – it becomes about containment and disaster recovery, both from a technical and a human perspective. These plans get put in place for contingency and back up.

Grief can often sit in our systems as dwell time – remember that this can be a very long time – then suddenly, a trigger can happen, and the grief will attack us out of nowhere. At this point, like a cyber attack, we lose control over our rational minds and can easily allow the same emotions, such as fear, shame, guilt and worry, to control our lives and hijack our behaviours.

To manage this, we can use the same principles that James speaks about in terms of communication and contingency. Being honest, open and transparent about how you are feeling and communicating this as best you can to the people around you can be a huge freedom and help the emotions dissipate so they don't feel so overwhelming.

For example, you can have a go-to person that you can speak to when you are feeling low. If you don't feel like getting out of bed, there can be someone on hand that you can ask for help, to cook you a meal or to spend some time with you. You can join a support group to help you. There are a few contingencies you can put into place that will really help you remember that this attack is non-malicious and will pass.

Trust and transparency are so crucial in our lives. Self-awareness and the ability to plan our lives in this way is a nice idea. On paper these ideas may seem obvious, but the challenge is that grief attacks the long-term agency and intellectual control centre in the brain. This is why education and empowering yourself with information are so important. The risk assessment exercises, the fire drills, the roleplay of disaster – these practices happen in all sorts of industries for people's safety and for damage limitation. Firefighters run up and down training towers in between putting out real fires because the constant practice keeps the brain in check, it keeps the mind responsive and helps eliminate stress-related shutdown or panic. The drilling protocol is crucial.

We need to practise keeping our own emotions under control and this takes a lot of self-enquiry and self-discipline. We all need to have a grief protocol in place. The grief protocol contingency can be as simple as a checklist or a list of steps and has to include asking for help.

Once you have established your very own grief protocol and you have it firmly in place – and this can also include a rule book and a metaphorical toolkit to use – then when grief strikes and hijacks your emotions, you have always got your protocol to lean on.

There is no shame in asking for help. Remember one of the ways a hacker can get to you is by using your sense of shame and embarrassment. The same is true when grief tries to hack you, particularly when you feel that you should have 'accepted' the loss. In such situations, we are often too ashamed to say that we are overwhelmed with grief as the rest of the world appears to have moved on, but this shame simply compounds the power of the hijack.

There is a nasty attack called the romance scare. I need to share this, again, not because I want to scare you, but I do really believe that education is power. Once you know something, you can do something about it. If you don't know it, you can fall for it.

Interestingly, James does not like the phrase 'fall for this' – he uses the term 'caught up in it'. He says that he does not want to say you fall for it, because a lot of these scammers are so highly manipulative: they know exactly what they are doing and you have been targeted and screened so carefully that it is not your fault. There is nothing about your behaviour that you need to question and doubt, because you simply got caught up.

The romance scare relies on the perpetrator setting up a fake profile on a dating site and finding the most vulnerable targets there: the widow, the divorcee, the attention seeker. They use your brain chemistry against you, triggering reward and love by saying and doing things that release dopamine and oxytocin (so-called happy hormones) to make you fall in love with them. What you are falling in love with is an idea, a promise, a hope. It's horrible.

The social-engineering scammer will know how to tap into the hearts of these vulnerable people and will make promises, woo, give compliments and manipulate emotions to make their target fall in love with them. They will be charming, flattering, say the right things, give them attention – give them what they are lacking emotionally. The scammer will seemingly build up their target's self-esteem by attaching themselves to that feeling. And once love is involved, things can get very messy.

After a relationship like this is established and built, it quickly becomes filled with control and manipulation. 'I say jump, you say how high. I say I need money, you say how much. I say I need you to sign this document, you sign.' Small requests at first, tokens of kindness, lead to more impactful requests: 'Oh my goodness, I've just rewritten my will.'

The truly scary fact is that this is so common, it happens all the time, and it can go on for years and years. We don't hear of it as much as we should because the victim doesn't want to be named and shamed, and the shame game continues its evil cycle.

If you can learn to override these emotions with intelligence and the right support in place, then you will lessen the likelihood of being attacked and manipulated in this way.

Dating sites, I think, have a moral responsibility to take down fake profiles. However, it's not always easy to spot them. If you use a dating site, you naturally lean towards the profiles of people you like the 'look' of – attractive supposed singletons who appear successful, kind, loving and supportive. Fake profiles have good-quality pictures and are mini advertising campaigns. Good looking, good job, interesting hobbies. Maybe a picture with a dog, or a cute-looking outfit. Perhaps a bit of skin exposed … You get the idea.

People get caught up in this idea of lust and trust. The scammer will know how to use lust and desire to get you hook, line and sinker. Like when phishing/fishing, the scammer baits the hook, and waits. There is a sequence of events that has to happen. The bait is the promise of an emotional need being met. This emotional need does get met and through that you are hooked.

Then the reeling out is the test of 'faith' – being asked for a small thing, and that small thing is given, because you want to help or please or remain in favour with this person you have fallen in love with. You may even face a moral dilemma, as not acquiescing will play into your fears of rejection and missing out. You will feel bad if you say no. So, you say yes. Ten pounds leads to a plane ticket. Oh, but then when they don't turn up, it's because a disaster has happened at their end, and they may need some of your generous understanding and sympathy.

At that point, it will hurt you more to doubt them, because they have been so nice to you. Doubting them will highlight your own naivety, and this is how we can so easily go into denial. No one believes that someone would do that to them, because they would never do this to someone themselves. You grieve the idea of love and justice – the values that you believe the world is founded on, and this grief at realising that this isn't always true can be so hard to face that you'd rather just keep the illusion up.

Grief can often show up when we are placed in situations that don't feel comfortable and are at odds with our innate values and beliefs. Sometimes this can be because we are trying to make our loved ones happy or to fit into a tribe where we simply don't belong, or it can be an adherence to a religion, or at the more extreme end a cult, that requires us to deny certain ideas about who we are in order to fit with that belief system.

In my research for this book, I interviewed Carolina (not her real name), who had been born into a strict religious family and, as a child, experienced abuse from within the church community, and in later life denounced her faith. She shared with me how, for her, the dominating power of individuals in leadership roles within the religious institution meant she lost her own sense of autonomy and sense of what was right, as she'd been brought up to not question what was being asked of her, because she'd been told it was what God wanted.

More than that, she believed implicitly that if she adhered firmly to the rules of the religion, she would be rewarded in an earthly para-dise after she died, along with her loved ones, which gave her this incredible sense of belonging. However, it was also predicated on the fact that she couldn't belong anywhere else, other than within that belief circle. On leaving the faith, she became entirely estranged from many of her closest family members and, beyond that, all her support network that was so bound up in the religion.

Crucially, she had also lost everything that her life had been built on; all her belief systems about what was good and bad had been shattered. Such a complete disconnect from the identity that had been built for her triggered intense feelings of loss, and her way of channelling these was to establish a new life based on her innate values, those beyond religious doctrine. But her experience has left her acutely aware that in situations where you are encour-aged to cede your own identity to a collective power or belief, you are extremely vulnerable, and she now works as an advocate for the safeguarding of children.

Carolina's story is a tragic one in which she has been left grieving her loss of self, her childhood and her trust in those around her, the institutions she'd been taught to believe in and even in some of her own family members. There is no simple response to this, but it reminds us how important it is to do what Carolina did, which is to hold true to that inner voice, our core identities, that tells us when something is wrong, as this instinct will help us find a way forward that is right for us.

## IN GRIEF AND GRATITUDE

Many of us experience grief when the idea of something turns out to be at odds with the reality, which can happen in even the most seemingly happy situations. Marriage is one very complex example of this. Marriage is about a coming together of two people who love each other and choose to spend the rest of their lives together, but it also comes with a ceding of control – a loss of absolute autonomy in the move from a single life to a shared one.

At the start of the marriage journey, we think about the happiness of being proposed to and the excitement of being presented with a gorgeous diamond ring, coupled with the feeling of being wanted that arises when someone gets down on one knee and declares their love to you in such a thoughtful way. We hear about the fairy-tale stories of these wonderful occasions, but there is another side to this, and it's important to share this, because even with good things, there can still be grief.

India's story is a fictionalised account based on the many experiences that people come to me with around marriage and the huge identity shifts it can trigger. India started having hypnotherapy around six months before her wedding day. She had turned 30 the year before and for her birthday her partner Michael had whisked her away to New York as a surprise. She was delighted as she had never been there before and had always wanted to go.

He had another surprise up his sleeve. While they were there and on her actual birthday, they had planned a trip to hop across on the ferry to the Statue of Liberty island.

She recalls being on the ferry and looking up at the statue getting closer. She was reading all about the history of why the statue was erected and began thinking about the idea of freedom and what liberty as a concept meant to her as a person. They arrived on the island and it was a windy day, but warm. They walked around casually, holding hands, and everything felt easy and relaxed.

Michael and India had been seeing each other for a few years, and he had mentioned getting married in the past, and she had always brushed it off; she never felt the need to cement their lives together in this way, and felt that marriage was an institution that would tie her down and limit her life. (Notice the language she was using surrounding the idea of marriage.)

You can probably guess what happened next. This is the part where, if this were a romantic film, the peaceful piano music would stop and the record scratch would rudely interrupt this potentially perfect love scene.

While they were walking around, Michael got his phone out and began speaking into his video. He pulled India into the shot, and she realised that he was using Facebook Live. Then, Michael did what a lot of ladies would probably dream of – in front of however many people watching this Facebook Live video, he proposed to her.

She describes this paradoxical moment so gloriously. She told me that, inside, she felt like a human-shaped hole had been suddenly punched into the ground beneath her, which she immediately dropped into. She described it as being suffocated, as if she were being buried alive. She couldn't breathe, she couldn't speak. Her focus became tunnelled vision, and all she could think about was finding a way out of this hole that she had been sucked into and getting off the island. Tears fell from her eyes, and they were tears of fear.

On the outside, for the viewer of this unfolding film, she managed to convert her internal panic into an Oscar-winning public performance: she appeared genuinely and flawlessly smiley as she feigned delighted surprise, and everyone, including Michael, were convinced those tears were of joy.

She tells me that breathlessly out loud she said, 'Yes', while every cell in her body was screaming, 'No'.

There was a cheer from the phone, and Michael gestured goodbye to the watchers on Facebook, and as soon as he put his phone away, he hugged her, and they both dropped to the ground as she burst into tears and sat on the floor, looking over his shoulder up at the huge Statue of Liberty in front of her as she felt her own sense of freedom and liberation immediately shrivel up and disintegrate into dust that blew away across the water on that warm, windy day.

Michael sat with her, holding her for what she says seemed like a very long time, completely oblivious to the fact that she was crying with grief, rather than gratitude.

This is such a good example of how an experience that we may all assume should be happy and filled with joy may be the opposite. Grief is a funny feeling that can be easily mixed into the cocktail of happiness, and it can create an interesting flavour – which is an acquired taste.

The taste of grief is something that I ask my clients to get familiar with. While working with India, I explained how she was grieving parts of her life that hadn't even happened yet, and grieving her own sense of who she was, her identity and her idea of what freedom meant to her.

Over the weeks, she began to alter her ideas of marriage being an institution and a trap; she began to change her language around it and started to call it a forever friendship, a partnership. She started to talk about how she could imagine enjoying the sense of solidarity they would be entering into together and began to look forward to uniting with her soul mate.

This was a beautiful transformational reframe, and one which I am grateful to have witnessed. She was letting go of the belief systems she felt were holding her back and preventing her from enjoying their relationship and the developments and growth that she knew she wanted, and she was updating her ideas about marriage in a way that still felt totally congruent to her. She changed her opinion of marriage from being 'an institutional trap' to a 'loving union of hearts and minds'.

They chose to have their wedding rings made in copper, because the Statue of Liberty is made of copper, and the reason it looks blue is because it has oxidised. The oxidation of the rings, she told me, will give them both a nice reminder of things being given the freedom and flexibility to change and to age together gracefully.

India sent me a card when she and Michael reached their five-year wedding anniversary. It had the Statue of Liberty inscription, a poem called 'The New Colossus' by Emma Lazarus, on it, and India noted that those powerful words remind them both of their choice to unite in their forever friendship freely.

Having found out about the history of this iconic global emblem, I'm sure this same sentiment was at the heart of France's gift to America:

'Keep, ancient lands, your storied pomp!' cries she
With silent lips. 'Give me your tired, your poor,
Your huddled masses yearning to breathe free,
The wretched refuse of your teeming shore.
Send these, the homeless, tempest-tost to me,
I lift my lamp beside the golden door!'

Even if we may not have a strong objection to marriage, and we find ourselves walking down that aisle, each step forwards is a step away from unattached life. Each step towards that person you are marrying is an agreement that you will abide by the vows you make; each step away from the door is a step away from a life of complete independence and volition.

For a lot of people, this is exactly what they want, and the idea of partnership, companionship and togetherness is one of comfort and utter reassurance. We, as humans, work better as a tribe than individuals, but really we must also honour the idea that signing away some aspects of our freedom and liberty is still relevant on so many levels. The sense of two become one means that the one we lose is ourselves, and that is something we should recognise and acknowledge.

India started her sessions by saying that she didn't want to feel like she would be signing her life away, and she wanted to get her head around the idea of marriage being a positive thing, and hoped the sessions would help her be grateful and happy about her choice to say yes. She knew that she did genuinely want to be with Michael forever, and that she was sure she wanted to make this commitment to him, but she wanted to get rid of her fear and anxiety around this idea of commitment. She understood that she needed to make this shift in her thinking and that was why she chose to come and see me.

We worked together for six months and by her wedding day, she was happy to sign and speak her heartfelt pledge to Michael: 'I honour this loyalty to Michael, the man who stands solidly, like a statue by my side, as I will stand by his, as his lady. May we both continue to fill our forever friendship with adventure, joy and liberty and stand together in freedom ...'

## FOR RICHER, FOR POORER: BUSINESS GRIEF

The marriage vows 'for better, for worse, for richer, for poorer' can extend out from marriage to other ventures and experiences in people's lives.

Essentially what we are vowing to do is commit to something in a wholehearted way even when things get bumpy and difficult. When we take on this sort of commitment – whether it is to become a parent, join a community or religion, or start up a new business – we are jumping into something with our whole bodies and minds, and that thing becomes our business, and that business (whatever it may be) becomes personal.

This is, in fact, the message that Penny Power OBE, author of *Business is Personal*, shared with me in an interview for this book. Penny, along with her husband Thomas, found themselves broken after their trailblazing business collapsed.

They were the first to market with their venture. It was a new idea. In 1998, when the internet was just starting up, we had Six Degrees, but there was no online platform for businesses to network. These were the days before the modern-day online giants of Facebook, Twitter, Instagram and LinkedIn.

The only real-life business-networking opportunities were joining the Chamber of Commerce, the Institute of Directors or the BNI (Business Network International) – the offline networks through which business owners and entrepreneurs would meet for breakfast, lunch or after-work networking.

Thomas is an amazing connector of people and Penny describes herself as an emotional problem solver who can see gaps and knows instinctively how to fill them with a solution. Back then, Penny suggested to Thomas that she would like to start an online and offline business-networking platform that was supportive and collaborative. Their ethos was 'friendship first, business second'.

It was a smart, incredibly progressive and very brave idea, which Thomas loved, and so the Ecademy was born.

A sophisticated website was created and built – one that supported brand-new technological functionality for the late nineties, such as membership administration, subscription services, profile pages, internal messaging and meet-up groups, all of which we perhaps take for granted today.

At this point, Penny was mum to three children all aged under 5. Thomas was travelling the world, talking about this thing called the 'internet' and, through his passion and love of connectivity, he was able to spread the word far and wide, and he got the interest for Ecademy going.

It wasn't long before they had 1,000 members signed up, all paying either £10, $10 or €10 a month to network with like-minded people around the country and, because of the internet, around the world.

The foundation of the business was based on support and assistance, and thus Thomas and Penny nurtured their fourth child into a very successful enterprise. They made sure that the culture of Ecademy was always friendship first, and got to know every single one of their members very well.

Penny and Thomas became the 'power couple' who had built this pioneering networking business and, at its peak, Ecademy had over 650,000 members.

Ecademy grew exponentially in members and in good will, and as it grew over time in this powerful way, it needed investment prior to being floated on the stock market. They raised £750,000 to show that they had serious investors in place and were given a date that they were due to go on the stock market of 18 March 2000.

Thomas folded his other businesses to put all his money, energy and time into Ecademy and, as a family, they put all of their faith and possessions into the business. They were personally valued at £22 million. Life was good.

However, Ecademy was not actually providing them with any personal revenue – maybe a bit of sponsorship and the consulting revenue it was generating, but much of it was being ploughed back into the now hungry mouth of the asset-driven Ecademy.

They were seventeenth in line to get on the stock market when the 2000 financial crash happened. The stock market shattered, every dot-com business on the stock market shattered, and Penny and Thomas' world shattered as a lot of their members lost their businesses. Ecademy as they knew it then was potentially over, just like that.

One of their major investors wanted to fold the business and turn it into an e-learning platform. This didn't sit right with Thomas and Penny. They were passionate about networking, building relationships, making friends in business and helping each other. The investor pulled out.

This financial hit was hard, but Thomas and Penny were determined to save their business from sinking. They began propping it up with more money, more time, credit cards, loans and goodwill.

They started a new type of membership service with a subscription model, which is a common feature nowadays, but back then it was a very new concept. The subscription service saved Ecademy, and Penny and Thomas' business continued to grow across the world.

In 2002, one of the members of Ecademy, Reid Hoffman, also saw the potential and massive power of digital networking. He managed to raise $330 million to plough into a new start-up. LinkedIn was born.

In 2004 Facebook was founded, a platform originally started for Harvard students to keep in touch with each other, which then mushroomed into the global social platform we know today.

At first, the emergence of these new platforms didn't have a huge impact on Ecademy. However, over time, Ecademy's membership numbers began dropping. There was a sense of 'why do we have to pay for something that we can use for free?'

On a business level people could use the LinkedIn networking tools and, on a social-networking level, there was now Facebook.

So, Thomas and Penny were being challenged to make Ecademy a free-to-use platform. Thomas and Penny were wary of this because of their concerns around the other implications this might have in terms of needing to monetise their members' data. Thomas and Penny were not prepared to compromise on their personal values of treating their members as human beings and not statistics.

Slowly, slowly, slowly the free internet won; however, incredibly, Penny and Thomas carried on going and growing, for a few more years, with one lifeline left.

One more strike, and that was it: they were dealt their final blow on 20 July 2012, as the banks faced a huge crisis and demanded immediate payment. There was no more choice left but to let it go. Ecademy, and all its liabilities, was sold for £1.

Thomas and Penny's fourth 'child' had died at just 14 years old. They had invested their love, attention, energy and time into nurturing this entity into its teenage years, when they lost it, just like that.

Penny describes her grieving period as being broken. Broken financially, emotionally and spiritually. Heartbroken.

Their identity was shattered, the family was devastated, and they lost their home, their stability and their way of life.

Ecademy was not just a business – it was their life, their reason to wake up in the morning, their passion, their purpose. They connected deeply with their members; they knew about each one of them. Penny and Thomas' members became friends who were involved in their lives and watched their children grow up.

Penny and Thomas had carefully and painstakingly created a heart-centred business model based on ethical principles of friendship, kindness and respect – a place where people could support, share and care about each other. This networking circle was founded on friendly connection and this built trust and won business.

Humans need humans. We know what it's like to be starved of human contact and interaction because of what happened to all of us in 2020/21. Most humans love networking and building relationships. We love to support people we like. This is a beautiful way to be in business. To help each other thrive, not just simply survive.

So, they were grieving not just the business as an entity, but also the values that were at the heart of their business. When Ecademy was over, Thomas and Penny were emotionally crushed, but certainly not defeated.

Penny describes her family network as being the foundation for her healing and recovery. Even through all the house moving and the disruption, they all still had each other as a close-knit family for nourishment and nurture. The love and connection they all had as a unit became even more necessary and even more critical. They all became closer.

This strength to keep going was like a massive chemical boost of adrenaline, cortisol and histamine – the same chemistry you need to run away from the tiger that is chasing you. The chemistry you need to stay alive. This was their fuel. Their tanks got filled up with grief and actually this meant they still had some fight left in them. They now were boosted up with rocket fuel. Rocket fuel grief.

Penny went on to use her rocket fuel grief in an amazing way to propel her into another business. She very quickly launched the Digital Youth Academy.

This shift in energy and outpouring of her grief into her new business took its toll on her, but again, the business had grown into a massive enterprise. It became life-changing for the youngsters she was helping, and because of this incredible rocket fuel-like energy that Penny was firing off something amazing happened. In 2013, Penny was awarded an OBE. In her book *Business is Personal* (which I highly recommend), Penny describes it so beautifully:

> It was a spark of utter joy. The shock of this was enormous and its impact was life changing for me. I had a warmth inside me that I will never lose. At last we had been recognised for what we had done. Ecademy was no longer part of our lives, but what we had done had been acknowledged at the highest possible level. The validation and sense of self-worth was priceless. The Queen 'had packed my parachute'.

Losing a business, especially one that you have invested your time, energy and money in, and made huge sacrifices for, is huge. The sense of deep loss as well as loss of identity and purpose is massive.

Penny and Thomas learnt an important lesson from this blow, and it was about building a personal brand. A personal brand is yours, and you get to position yourself in a strategic place where you can be known for being helpful, reliable, honest and good. This is how to build goodwill, and Penny and Thomas had certainly done this. Nothing can take your personal brand away while you are alive, because it is you. With their overarching mission of connecting human beings together in a heart-centred way, unbeknown to them, Thomas and Penny were building a personal brand as super-networkers. The amount of human goodwill and karma points they amassed saw them through this huge chasm of business grief.

Please don't ever underestimate kindness – it really does go a very long way. Even in this age of doom and gloom and scary stories of manipulation and abuse, there is an innate nature in most human beings to be helpful, compassionate and kind. There are more good people on this planet than bad. That is a fact.

Penny tells me that she didn't have time to grieve the death of Ecademy, but I believe that she did grieve. Maybe she didn't realise it, but she used her grief to start up another profoundly impactful business and very quickly recover some of her identity, sanity and sense of purpose.

Grief isn't always about curling up quietly in a dark corner and rocking yourself into a depression, or crying constantly, or having dramatic outbursts of emotion at unexpected times. It can also be hugely transformational, like in Penny's case – a channelling of this magical power that fuels you to change direction, grow some new wings and soar higher.

# OUR SHARED GRIEFS

There are approximately 78 million people on our planet who will never, ever feel grief.

## WITH OR WITHOUT GRIEF?

These people who can't grieve are known as the 'one per cents'. Psychopaths.

Psychopaths do not grieve, because they can't. According to scientific studies, psychopaths make up 1 per cent of the planet. They don't know how to grieve because they do not *feel* in the same way as a neurotypical person.

When we think of a typical psychopath, we may think of the dramatised depiction of the serial killer, bunny boiler and Hitchcock's infamous shower scene. Recently, I binge-watched *Killing Eve*, and weirdly, the central psychopathic character Villanelle, a cold-blooded assassin, was enormously likeable and utterly charming. This glossy depiction is very similar to the character of Dexter – a vigilante serial killer – who is, again, rather likeable.

Psychopaths are not necessarily dangerous people like Hollywood would have us believe – they are people that think differently, because they feel differently.

Athena Walker is a diagnosed psychopath, and in a radio interview with Jeremy Vine on BBC Radio 2 in 2017, she tells him that psychopathy is not a condition, but more a collection of traits. The traits she lists are ruthlessness, fearlessness, impulsiveness, self-confidence, focus, cool under pressure, mentally resilient, charisma and charm, no empathy, no regrets, no depression or anxiety, no self-doubt and no sadness. She describes herself as a socially aware psychopath, which means that she has learnt behaviours to make sure she fits into society and has practised socialised norms so on the outside it would appear she has feelings, but inside there are none and – according to her – that is not a problem.

This collection of traits is fascinating. If a psychopath cannot feel and doesn't mind not feeling compassion, but can train themselves to act as if they do feel compassion, isn't that interesting?

If they don't feel sadness or depression, they also have no capacity to feel happiness or love either. With that in mind, would we want to give up our feelings even if those feelings are uncomfortable?

This goes back to the idea of duality. On one hand, there is the psychopath who cannot feel, and on the other there is the empath who feels too deeply.

Unfortunately, the brain is also a binary place – it's either/or: either you feel all the emotions, or none. It doesn't seem possible to feel only the 'good' ones and not the 'bad' ones.

But what if that were possible?

As human beings we have the capacity to learn; therefore we can train ourselves to get our emotional balances just right. This would allow us to genuinely feel all our emotions deeply, while maintaining a healthy level of intellectual awareness. This will give us a perfect harmony of emotional intelligence in the grey area of the mind. The grey matter of the brain is where chemicals are produced. The brain is a simultaneously simple and complex chemistry set.

There seems to be an interesting neuroscientific debate about whether psychopaths don't produce enough of a chemical called oxytocin in the brain. Oxytocin is the chemical that is predominantly released when females give birth to help them bond and connect with the baby, also to help them instantly forget the pain or any negative experiences in their labours. Some theorise that a lack of oxytocin means a psychopath has no desire to get close to people, bond or connect. However, this theory is not scientifically conclusive. In some psychopaths, there is a higher level of oxytocin present in the bloodstream compared to neurotypical people; this has confused scientists and led to the further understanding that oxytocin isn't simply the chemical of love and tenderness, but also the chemical of protection. It increases aggression and territorial behaviour, which makes sense because this will help the new mother find strength to preserve the safety and ensure the survival of her newborn. To protect her baby, no matter what.

Oxytocin, despite often being referred to as such, is not just a love bubble chemical; it is also a 'DO NOT TOUCH' chemical that helps our species survive. It stakes a claim on what is ours and what we know to be right and true.

It would then make sense that some psychopaths have high levels of oxytocin, because they can be highly territorial. They create a space that is exclusively theirs; they are self-centred and usually introverted. They are private people, with the capacity to distance and detach themselves to the extreme – and this explains why psychopathy is also called anti-social personality disorder.

Dan Baxter, a psychopathic researcher and psychopath himself, says in an online thread that psychopaths have a reduced emotional range and radar, and they also have mirror neuron issues, which impact their ability to empathise with others.

As this book is about how to deal with grief, the answer seems to be quite simple – but perhaps not that easy. If we all had access to our psychopath modes, and could turn our emotions on and off as necessary, wouldn't that be the answer to all our problems?

As human beings, we need to come up with a way to regulate our emotions better. Like an amp on a graphic equaliser, but instead of the usual things you see on an amp, like the volume, bass or treble, each slider is for each emotion instead. So, with the slider metaphor installed, we will be able to control our emotional levels much better.

If we could figure out how to switch the emotions on and off on demand with the ability to dial them up and dial them down – that would make life a lot more straightforward.

However, unless you are one of the 1 per cent, then it is a near certainty that during your life, you will grieve (at least once) for these three reasons:

1  You miss something and feel loss
2  You want something back from your past
3  You long for something to be different

Grief is an emotional cocktail of intense feelings, ranging from sadness, regret and anger to loneliness, fear, guilt and disbelief.

It can also be an absence of feeling – a longing, an emptiness, a feeling of being lost coupled with homesickness. There is a hollow incompleteness that needs to be filled. It's an uncomfortable vacuum, a void that we want to avoid.

This book is called *Planet Grief* for a very simple reason – apart from the 1 per cent who can't grieve, the rest of us on the planet will spend quite a lot of our lives grieving, and a lot of people may not have realised that the feeling I've described here was even called grief.

Planet Earth is an interesting place. It is a place of duality: north and south, east and west, night and day, light and dark, right and wrong, left and right, up and down, back and front, positive and negative, alive or dead.

Yes, there are shades of grey, and these shades sit within the spectrum of the extremes. The extreme ends are the sides of the container for the shades to sit in. We must have the extremes for everything to exist in between.

If we go with this extreme idea for a moment, you may like to think about grief as the thought of wanting to be in the opposite place from the one you are in. If you hate the winter months and are longing for the days to get longer and lighter, you are grieving the sunlight.

Grief doesn't always have to be loud and invasive, so you may not be displaying intense emotional reactions – it can be a quiet feeling too, an underlying discomfort, a generalised feeling of being out of sorts.

I think that grief plays a central role in the makeup of many mental health disorders – like seasonal affective disorder, postnatal depression, imposter syndrome and body dysmorphia.

Each of the aforementioned disorders have similar traits: a sudden change coupled with a dislike of what is being presented. This can create confusion, denial, uncertainty, ambiguity and a feeling of lost control – the same symptoms that my clients report when they are grieving.

When something has suddenly and significantly changed in our lives, and before we have had a chance to fully process, assimilate and adapt, in that period of adjustment sits the unknown and the uncertain. This feeling is what I am calling grief. Grief is the opposite of acceptance.

## GRIEF: THE PUBLIC TO THE PERSONAL

Can you remember what you were doing when you heard the news that Diana, Princess of Wales had died in a car crash?

I can. Everything and everyone stopped what they were doing to listen to the radio, to turn on the news. It was 31 August 1997: there was no internet like we have now, there was no rolling news, there were no smartphones.

I had just started working as a runner at the BBC, and I remember walking into the huge glass foyer of Television Centre on the Monday feeling exhausted because we had spent the whole of Sunday watching the news and the day had somehow merged into a night of little sleep.

When I walked into the office, everyone was crowded around the small monitor in the corner of the room. Nobody was sitting at their desks. Some people were sobbing; even some of the senior management were crying. Some people were completely stony-faced, almost zombie like. The atmosphere was tenderly sad, silently surreal and there was a very strong feeling of shared connection.

Some of my colleagues were hugging, some needed propping up, some were literally standing with arms closely linked or their arms around each other as they hung on every word coming from the TV.

The shocking death of the People's Princess had triggered a wide range of emotions, just in our small office. It was certainly an eye-opener for me as I watched how my colleagues acted over the course of the day.

The beginning of the day was very quiet – hushed tones, the occasional sobbing. By lunchtime, there was much chatter – bargaining and confusion – as people were analysing and breaking down what we had all been hearing. It was obvious that there was not much information being distributed – it was just the same information being repeated in different ways.

One floor below us, we had access to the BBC News Desk and one of our senior managers was reporting to us the latest updates from the news editor, which were nothing much. There was so much speculation by lunchtime as everyone tried to make sense of the information, desperately looking for answers.

When more information began to drip feed through, and we found out about the paparazzi, that was when the office got heated. The voices got louder, the opinions more vocal and the morning's gentle simmering turned into a boiling anger. Anger mixed with disbelief. There was also a lot of guilt. Guilt for working in the TV industry. *Were we to blame? Was this our fault?*

Our department was grieving, the whole floor was grieving, all the people in the Television Centre building were grieving. The whole of the BBC was grieving.

And nobody, not one of us, as far as I was aware, apart from the royal correspondents, had even met the woman. We were all grieving for someone we had all just seen on TV and read about in the magazines and newspapers.

We were also grieving for her two little boys who had just lost their mother. The princes were not shown on TV at all that week, as far as I can remember; we all could imagine their little faces, but couldn't imagine how they must have been feeling. We were all sharing their pain and their loss. We were all grieving for them and with them. It was incredibly powerful.

It was a week that will forever stay etched in my memory as a collective grieving process.

We all came close to that collective type of grieving again very sadly two years later, in 1999. This time it was for someone who we had seen, somehow been connected to, and had either spoken to or at the very least smiled at. Jill Dando.

Jill was a much-loved BBC newsreader and presenter. She had been given the title of BBC Personality of the Year the year Princess Diana died and, for us, she was our very own princess. She was loved by everyone she met; she lit up the room with her smile and her wonderful laugh. She was extremely down to earth and an incredibly kind and lovely person.

I had been working on *Crimewatch* for a short period and got to know her quite well while she was recording all her voiceovers for the series. It was my job to make sure the dubbing studio had the right resources booked in.

When I found out that she had been murdered on her own West London doorstep, a cold chill travelled down my spine, and this was the catalyst for my own personal breakdown.

This is the moment in my life when time stopped ticking. It felt like one day I woke up and I had lost any sense of reference about who I was, what I was meant to be doing and where I belonged in the natural order of my life. Grief came to find me, muscled its way into my system and, like a parasite, began eating away at my thoughts, my happiness and my emotions. I was being eroded from the inside out.

It was a hot day in May 1998 when the phone rang at work. It was the year in between Princess Diana and Jill Dando dying. The phone rang and, on its small dot-matrix display, the words EXTERNAL CALL flashed up. I never got any external calls, as all my friends and family had my mobile number and not my work number. While I was busy watching the words flash, I missed the call and it went to voicemail.

There was a tiny part of my brain that did not want to pick up that phone. The only time I had given my work phone number to anybody was when I'd listed it at the hospital that my dad was in at the time, with the condition to only call the number in an emergency.

I waited for a few seconds, and then the voicemail alert began flashing, and my heart sank into the pit of my stomach. I knew this call was bad news.

I was right. It was a consultant calling from Hammersmith Hospital asking me to come in as soon as I could. I dropped everything and made my way there. Within the hour, I found myself sitting in the stuffy consultant's office. He broke it to me nice and gently, but the long and short of it was that my dad was in the advanced stages of leukaemia, and there was nothing they could do for him other than keep him 'comfortable', and that his best hope was six months and, worst-case scenario, three months.

They were bang on. It was three and a half months. September 1998 was when my dad died, and this was my first experience of grief.

It came swinging in like a bulldozer and knocked the stuffing out of me. It came out of nowhere, a year later, one morning as I sat at my desk, just watching the footage come in about the death of Jill Dando. Up to that point, I had buried any feelings about my dad's death, but that day the trigger had been pulled as I watched the news roll in about Jill.

By then News24 was on air and we had the rolling news covering the whole story. My desk faced the TV screen, and the constant talk of death and loss sparked something inside me, and I felt like I was on fire. My brain's fire alarm started ringing. It got louder and louder, and I knew I had to take this seriously and book some time off. I felt like a bomb had exploded inside me, and I needed to figure out how to put myself back together again. I realised that this grief thing was something I could not ignore.

My dad's sudden death was sandwiched in between the two princesses. All of the griefs shared the same qualities of disbelief, denial and shock – leading to bargaining, guilt, loneliness and a lot of moments of raging anger.

I felt like the collective grief of the nation was being poured in and out of me. It was a very intense time and it taught me just how powerful grief is.

The sense of a world in grief came again when the unthinkable happened at the Twin Towers in New York in 2001.

I was eight and a half months pregnant with my first son Krishan, sitting on the sofa watching daytime TV and eating my cheese and pickle sandwich, when the programme was interrupted by the news flash.

I remember watching the footage in total disbelief as the reporter was talking of a plane crashing into the building. While the cameras were still rolling, we all watched something that took our breath away. I dropped my sandwich and thought I was about to go into labour in that very moment as another plane flew straight into the second tower.

There was then a horrifying realisation that this was not a freak accident, but a deliberate attack – an act of terrorism – on the World Trade Center. What a moment. It was a sickening reminder, now imprinted on millions of minds around the globe, that our world is brutal. There are certain things that happen in our lifetimes and in life gone by that we cannot even begin to understand because we cannot even begin to think like that. World wars, genocide, brutal crimes and terrorism punctuate history and the grief they engender seeps into our collective consciousness and changes who we are and what we believe as humans for years to come.

When events such as the Twin Towers attack occur, we experience complete incomprehension, denial, disbelief and bargaining. Classic grief. We instantly began grieving all the precious lives that had been lost; later we discovered the death count was almost at 3,000 people, with approximately 6,000 injured, including just over 300 firefighters who died on duty. We also started grieving for thousands of families that had lost their loved ones and were affected by this unbelievable crime.

Like when Diana's death was announced, we can all remember exactly what we were doing. The memory of that moment is branded on to our brains. Forever.

The grieving continued for a very long time and continues to this day. Over 10 million people have visited the World Trade Center memorial and museum. The museum alone gets around 9,000 visitors every day. That is a huge testimonial to show that, even decades later, this is not just a museum. This is a monument of collective grief.

As well as the lives lost, we grieve those buildings: the 200,000 tonnes of steel that fell and 40,000 panes of glass that were shattered. They were part of the iconic New York skyline – a skyline that was instantly changed forever. The grief will remain in that gap, where those towers used to be.

Human beings have immense power – a force that can be used to wreak havoc or one that can be used to heal.

The same is true of grief. If we let it, it can destroy and crush us. However, if we harness it, it can strengthen and fuel us.

## OUR QUEEN IS GRIEVING

Princess Diana's death became such an immense global outpouring of grief that, at times, it felt as though the deeply personal nature of that grief in terms of her own family and friends had been forgotten. Only recently, we were reminded of that complex mix of national and family grief when the British media broke the news that Prince

Philip, the Duke of Edinburgh had died on 9 April 2021. He was married to the Queen for seventy-three years, and had dedicated his life to being her loyal consort. Another historic landmark moment. Queen Elizabeth II has been Britain's queen for my entire time on the planet, and now her husband, the Duke of Edinburgh, has passed away, just short of his 100th birthday.

Britain has just been coming out of the lockdown period facing the intense grief that that has brought and, again, we are reminded of the impermanence of our existence – as death is so much a major part of all our history and all our lives.

This royal departure has left many people experiencing 'imposter grief' – a feeling of grieving, but not for an actual person that they knew, but rather a figurehead of society – grieving a part of our history that has just died. Even though many of us never met or knew the Duke, he was still very much in our awareness, and we can still face the grief of this loss just as much as if we knew him.

I'm not a massive royalist, but the royal family have been such a huge part of my culture being a Londoner, and part of me feels extremely connected to this history.

A very small group was allowed to attend Prince Philip's funeral at St George's Chapel, Windsor Castle. Coronavirus restrictions meant the ceremony had to be hugely scaled down, with only thirty guests allowed, in contrast with the estimated 800 originally planned.

The thing that struck me so hard was that at the royal funeral, because of the same social-distancing rules that applied to all funerals conducted during the UK lockdown, the Queen was sitting all on her own, at least 20 metres away from everybody else. It brought home to me, and to many others, the acute isolation that everyone who has lost someone during the pandemic has been placed into, particularly during those rituals that would normally bring the bereaved into the circle of loved ones for comfort.

Here is a lady, the Queen, who has lost her loyal husband of seventy-three years. Theirs was certainly not your average relationship, as he was also her devoted and faithful consort, and this loss is massive. The image of her completely separated from the rest of her family in a black coat and black face mask, head bowed down, will stay with me forever.

Alongside the pictures from the funeral, there was another image displayed of her in the press with one solitary tear running down her cheek, and when I saw that, it completely broke my heart. Regardless of your wealth, status, or power, it doesn't matter who you are, you cannot escape death, loss and grief.

## DIVIDED OR UNITED BY GRIEF

Grief can materialise in the most unexpected places. Perhaps you wouldn't expect grief to strike in the echoed chambers of parliamentary Westminster, but it does and it's certainly incredibly topical as I type these words out on my laptop. Right at this very moment, Britain is leaving the European Union. It's Brexit Day! Great news for the people who voted to Leave, but not so great for those who wanted to Remain.

Pollster YouGov recently released a survey of Remain voters in the UK, in which it asked which of the Kübler-Ross five stages of grief (denial, anger, bargaining, depression and acceptance) most closely described how they feel about the EU referendum result. All of the feelings ranked highly, from depression, anger, denial, disbelief to bargaining, and they discovered that almost 70 per cent of Remainers are in deep grief.

Remain voters account for 48 per cent of the voters in the referendum, and because it's nice to put proper numbers to the percentage, that is approximately 16,141,241 people, and 70 per cent of that number is 11,298,869. That's a lot of grief. Interestingly, most of these people won't even realising they're grieving, but it's showing up in all sorts of ways.

When the 2016 Referendum results were announced, a lot of my clients began cancelling their sessions, with imminent concerns about their financial stability. This was interesting to observe, as nothing had actually changed for them that day, or that week, or even that month necessarily, but panic and confusion seemed to be in the air.

The uncertainty of what was going on politically was feeding through to the personal and people were questioning how this was going to affect their income, employability, living conditions and their standards of living. There was talk of the deepest recession ever with house prices devaluing and food becoming too expensive to afford.

There was a general feeling of loss of security, coupled with fear of the unknown. This was especially pronounced in those for whom the change clashed with their own belief systems; such a collision of events and values can shift people into a state of self-preservation and the protection mode in the brain gets switched on.

As a therapist, this is interesting for me to watch because I often talk to people about the brain being comprised of two parts, and in this situation I could quite clearly see my clients moving quite rapidly out of their intellectual and higher-order-based thinking and into primitive thinking.

The primitive and emotionally narrow-focused and protective part of the brain helps us ensure our safety and survival on the planet, so when something changes on this scale at a national level it's a bit like your parents suddenly announcing they are getting divorced.

If the parent – in this case the country that we live in – suddenly starts to fall apart we – as the child, the inhabitant of the country – can feel this sudden schism, and we start to wobble as we try to figure out which side of the divide to stay on and how best to find a sense of comfort in that choice. To protect ourselves from hurt or danger, we will go into fight, flight or freeze, or a combination of all three states. It's purely the brain trying to find its way back into homeostasis – which is balance.

The cave-dweller's take on comfort is usually to hide in the cave, to retreat, to run away, to bury their heads in the sand or to find some sort of release in the fight, the anger and the desire to protect themselves from the perceived imminent threat to their safety and existence.

Human beings can easily get lost in an idea which confirms their own beliefs, especially if they are being exposed to the idea often enough. Politics aims to do exactly that. Politicians will say the right thing at the right time to like-minded people. Those people will then go off and repeat those words, and a huge game of Chinese (in this case British) whispers ensues.

Throughout the Brexit debate, the newspapers, television, radio, magazines and social media feeds were all informing the debates around the dinner table, in the coffee shops, over the phone and online, and conversations, even while walking the dog on a quiet Cornish coastline, were loud and long. Leavers and Remainers were in instant conflict with each other. There was an instant divide – a wall. It became one vs the other.

Divided opinions make for better debate programmes. People who are happy to nicely agree with each other make boring programming, and this doesn't get great viewing figures. TV shows that encourage participants to pit themselves against each other and promote competition are the ones that trend.

I've sat on a breakfast TV sofa three times now, and I know exactly what they are looking for. The producer is told to find two people with opposing opinions on a certain subject and the six-minute discussion becomes a fierce verbal battle. Essentially a powerplay of words.

This debate, if heated enough, will then make it into the tabloid press that morning or the next day and keep the viewing figures high as the publicity that the programme gets is fuelled by the human need to take sides. Sitting on the fence in these debates is not allowed. You must be on one side or the other. That's how it works.

Tribal society works like this. Tribes often stick together and everyone in the group will work together as a close-knit unit. Humans work better as a tribe. In terms of evolution, it was much safer to be part of a tribe than to be an isolated wanderer carving your own pathway through the jungle. Without the support and collaboration of the tribe, you might find yourself in the mouth of the tiger, and that would not be ideal.

However, when our tribe turns out to be divided in itself, we can experience intense grief as we have to choose from within our own tribe in order to become part of the newly emerging groups.

Grief is a reminder to us that we, as human beings, need to belong somewhere – whether that is in a tribe, a family, a belief system, a community, a relationship, a career, a vocation, a home or even on a planet. When we face a loss of that belonging, grief stares back at us. We gulp down that feeling of homesickness.

# GRIEVING EARTH

NASA astronaut Peggy Whitson has spent 665 days in outer space; this is more time in space than any other US astronaut in history. From growing up on a small farm in Iowa's smallest town to living and working on the International Space Station, Peggy has called many places home, but if you ask her where her home is now, her answer is simply 'Planet Earth'. She has taken feeling homesick to another level!

Many astronauts have talked about the startling shift in perspective that happens on the International Space Station when viewing Earth 254 miles away.

They call this the Overview Effect. They speak of a profound and deep appreciation for this delicate turquoise ball spinning in the black void of space, which from afar simply looks like a giant rock just hanging in the darkness within its silvery orb-like atmosphere, and then of a realisation that this is, in fact, their very own home – our home.

In *The Orbital Perspective*, NASA astronaut Ron Garan recalls the bittersweet feeling of viewing our fragile Earth from this planetary perspective and realising the innate paradox of being intensely grateful for a world that is protected from the 'harshness of space', but which has proven incapable thus far of protecting itself from the inequities of human division that have plagued its history.

This huge contrast in perspective speaks of the ultimate planetary grief – grief for what is and what could be. Still, Ron points beautifully to the hope that can exist within grief: 'Part of this is the realization that we are all traveling together on the planet and that if we all looked at the world from that perspective we would see that nothing is impossible.'

What if you could imagine looking through the eyes and minds of Peggy and Ron, and all the other astronauts who have had this incredibly powerful vantage point of our planet? Would this change you? Would you accept that grief, like death, is part of existence? Often to emerge from a grief state, a shift in perspective is what is needed. Sometimes, we as human beings can get so bogged down in the minutiae of life that we lose sight of the wider perspective and the bigger picture. We lose the ability to stand back and view the world from a bird's-eye view. Sometimes, it would be hugely beneficial if we could all experience this awe-inspiring feeling of the Overview Effect.

Greta Thunberg, in her iconic speech at the World Economic Forum in Davos, finished off her address to the congress with these words: 'Our house is still on fire. Your inaction is fuelling the flames by the hour. And we are telling you to act as if you loved your children above all else. Thank you.'

Human beings have the capacity to use their minds and collective power to make changes and make a difference. Small acts of kindness and thought can go a very long way to conserving and preserving our world.

# MOVING ON ...

As we have ascertained, we humans like to know who we are, where we fit in and where we have come from. This knowledge of our existence shapes our worlds and forms our identities. Our identity is very important for us to feel secure and solid. Identity gives us the sense that we matter – that we have a place in this world – and this makes us feel safe.

It can be shaped in a few different ways, but I want to explore how our parents start this process as we grow up, and how damaging it can be when the identity we thought we had gets taken away from us.

## MELISSA'S STORY

Melissa (not her real name) is a 56-year-old psychotherapist who has still not recovered from what happened to her eighteen months ago. In 2018, she got heavily involved with creating her family tree. She explains what happened next so beautifully – I will let her tell you her story.

> I had a new hobby of creating my family tree online. I got quite involved with it because one can create a tree and begin to fill it in as

you get more information. You start filling it in with the things you know – your mum, dad, grandfather, grandmother and it fans out and it goes back a long way. I got a bit addicted to this. I would do it every night as my hobby. I was concentrating on my dad's side because I've known very few of my dad's side of the family. So, I thought that was a good place to start exploring.

I spent months on it. On the website, you can contact other people that have the same relatives as you in their trees. I found someone on there, a woman that had the same great-grandparent in their tree. So, I messaged her, saying, 'Oh, I think we're related. We've got the same great-grandfather.'

To my delight, she replied, 'It's my husband's tree I'm doing it for him. Yes, you must be second cousin to my husband. How amazing.' She said, 'I'll send you all our genealogy paperwork and get you in touch with some other cousins.' She then friended me on Facebook. They phoned me and they sent me all this paperwork to add to my tree of all my dad's side, who was a Rountree [names have been changed]. It turned out that the Rountrees all came from the same county and the woman explained how the family was mostly still in the same town that they'd been in for hundreds of years.

I thought, *Wow, this is amazing.* She sent me a photograph of her husband, and he looked just like my dad, it took me aback, I'd never seen anyone look like my dad. I loved my dad, and it brought up a lot of stuff for me, as my dad died 14 years ago. So to now see someone looking like him, it was overwhelmingly emotional, I nearly got in the car and drove to sit on this man's knee!

I was thrilled, and I replied to her immediately, 'My god, it's like seeing my dad again, you know?'

She wanted to be even more helpful and responded with a suggestion for me. She said, 'Well, we've all done a DNA test. If you do a DNA test, you'll find that it will help you with your tree, so you can find more second and third cousins with the same DNA as you, you see.'

She told me that her and her husband had done theirs. I was so excited, I told my partner and he bought me a test for my birthday present. I did the test and a few weeks later, I got my results through and I couldn't find any Rountrees that matched my DNA in that area. I was confused because she told me that all I had to do was enter my DNA details and it would start matching up, but this didn't happen. I messaged her again and asked if she was sure she had her husband's DNA switched on, and explained that I wasn't matching with them or any of the cousins. She said, yes, it was definitely switched on.

Then, the penny dropped. I realised that as this man looks so much like my dad, he must be related to my dad. It was me that wasn't related to my dad. That was the only other option.

So, I knew then, the man who all my life I had thought was my dad was not my real dad and I was in total shock. It was torture, but it began to make sense, after all I'd never looked anything like my dad. I had always wanted to look like my dad, because I loved him so much. I always thought he was such a good-looking man. It confused me why I could never see anything of him in me.

As you can imagine this threw up so many questions for me, questions that I had no idea who to ask without opening a can of worms. When I realised that my dad wasn't my dad, I felt devastated, in shock and alone.

I was 55 at the time and felt like I'd been deceived all my life. At the time, my mother was still alive but in her nineties. My mother and I had a very troubled relationship from early on. She's a very uncaring and frightening person, she's never been someone I could go to for help or to even talk to. I left home aged 16 as I needed to get away from her as soon as I could.

Now, as I was beginning to piece things together, I started to understand what had happened to me and why I had always felt lost growing up, I felt I was in the wrong place, the wrong family and that I didn't belong. This heartbreaking discovery started me on a journey – a journey into deep grief.

I sat with this knowledge for a month, I lost about a stone in weight. I couldn't eat, and I spent a long time looking in the mirror thinking, *Who am I?* I didn't even recognise my own face.

You base your identity on who your parents are, who your grand-parents were and you believe you are part of that history, and then when I realised that I wasn't part of any of that, I felt lost in the world. My history crumbled. I was in a bad place. I was crying all the time. I couldn't walk, my legs wouldn't work, it was as though they had turned into jelly for about a month. I couldn't even get up my own stairs. It was awful.

However, in time I became determined to find out the truth and did eventually uncover my birth father's identity and the way in which my mother covered up her affair. Learning all of this made me real-ise how my mother's treatment of me throughout my life had been fuelled by her shame and anger at my very existence. With that reali-sation comes the grief for what could have been if I'd been born into a different life.

At the moment, I still don't know who I am – I'm still working out my identity. I'm hoping to get to that place as me. As a different person to my parents. I know that my life is full of grief, and I guess I need to learn to love the grief.

I know that my great traits of being tenacious, curious and adven-turous come from my biological father as he wrote an autobiography and I see myself in his personality. My love of learning and need for culture, as well as my thirst for knowledge and love of history – which was the drive to find out where I came from (hence doing the family tree in the first place) – every one of these positive personality traits come from my biological father, and because of these very traits, this led me to this crazy discovery, and uncovering the truth from 55 years ago, even through all the rejection, the abuse, the door slams, the hate and the fury – I just kept on going, I kept my mission in my mind and I found my real father, just before he died.

As human beings, we need to make sense of things – all the gaps need to be filled with some kind of reality or some kind of understanding for us to feel whole and complete. If lots of gaps and holes remain, we feel unstable and incomplete.

The human mind automatically wants to make sense of the world. The brain is constantly looking for clues to solve issues and move through challenges; this is part of the innate survival instinct that wants to find ways to stay alive and ultimately thrive. We are constantly asking, 'What's this problem? How can I fix it?' In terms of grieving our identities and letting go of who we thought we were and moving on to fully embrace who we are and who we want to be, it isn't a straightforward process. We can spend our lives circling around this sense of identity and grieving over and over again for the different phases of loss when things change. This can be particularly the case when our sense of self has been so tied up with a specific role. The pain of 'empty nest syndrome' is often tied up with grieving for the hands-on motherhood role, and the same can be true when it comes to retirement – for some it is a shift that is joyfully embraced, for others it brings a profound sense of loss and rootlessness. This can be pronounced when we are forced to stop doing something we love due to circumstances beyond our control. I met with the Olympic swimmer Sharron Davies MBE who reflected on her sense of loss when she stepped away from her professional sporting career.

## SHARRON'S STORY

For me grief is the absence of something that has become the norm. Of course, we can feel grief when someone dies, but it's also very much felt when something ends in your life. Another way of grief showing up is when your dream ends and then there is nothing to replace that dream.

I felt an enormous loss when I stopped competing. Swimming was a massive part of my life and filled all my time and energy – swimming was my world, my everything, my direction. When this ended, I felt a huge loss. I know I felt grief.

Once I stopped competitive swimming, there was a whirlwind of emotions. Initially, I celebrated the fact that I no longer had to get up at 5 o'clock in the morning and didn't have someone telling me when to go to bed, or saying I mustn't do certain things and constantly monitoring my food and drink.

When competitive swimming ended for me, I can only describe it as a 'falling off the rails' phase and I went and did lots of things in excess that I hadn't been able to do previously. The irony was that after a while of being totally 'free', I realised that the person that I really am is the swimmer – the one that likes the structure and wants the simplicity of working hard and getting a faster time. I thrived when I had a target to perform to. I also enjoyed being surrounded by other athletes who were very like-minded.

In my competitive life, the people around me were focus tools for me, because they were all focused themselves. It was a tough world, but we were constantly coached to be tough and resilient. When it all ended, it was a huge culture shock. I was out in the world where everyone is fighting their own little corner, and I realised that there was not as much care or support in this new 'free' environment. I missed my teammates, I missed a very, very strict routine, I missed having very simple targets, that could be six months ahead. We were always training for the next thing – either trials or the championships for that year, and then ultimately, every four years there was an Olympics to train for.

Being a competitive athlete for your whole childhood and young adult life you are taught to constantly strive for the next thing, which in hindsight is very limiting. It has shown me, that this may not be balanced for our future athletes. So, what we're trying to do now with athletes is to help them to understand that it's not a good idea to have all your eggs in one basket, and we encourage them to have something else such as academia or an additional focus alongside their sport, which gives them something else to concentrate on when sport finishes. This is the advice I always give young athletes when I coach them. It's not just for when they stop competing, it's also because you never know what may happen in life. They could sustain an injury which would put a stop to their athletic career immediately, and it's so important that people can have something else to rely on as back up – to avoid falling into a pit of grief.

During lockdown, I had a serious bike accident and broke my hip. I have to say, the grief of my accident was quite tough. I've always been someone that's been very physical and very able; my body has always been my tool. It's also accepting that the body changes as we age. I am a very fit and able person, and even at 58, I'm in great shape – but I still find this ageing thing tough. I am grieving my younger abilities. I'm grieving the fact that I cannot do what I was doing 30 years ago, and I know that I'm never going to be able to do that ever again – that is tough to come to terms with.

Losing my mum was the sharpest, most difficult grief to come to terms with – because there was no fixing it. When I gave up my sport, I could find activities to replace it, but when you lose a family member, there is no way you can replace that person – there is always going to be a gaping hole. A mum-shaped hole that cannot be filled, ever. So, I have had to learn how to adapt my life to include that gaping hole. There's always this burning desire to pick the phone up and talk to my mum. But I know I can't. They say time is the healer, but time doesn't heal you. I think time enables the loss to become the norm. You can miss somebody and have grief live with you for the whole of your life, this is fine if you don't let it affect everything else in your life. I'm not sure that I'll ever stop missing my mum and I'm not sure that I ever want to get to a point where I do. I believe I'm very lucky to miss my mum because it means that she meant something to me. Instead of making it negative, I would encourage everyone to turn their loss around and make it a positive. I had my mum for fifty-five years of my life, and she was a wonderful mum. I miss her like crazy, but the fact I had 55 years of my loving mum in my life – makes me feel very grateful.

Sharron tells her story so powerfully and it reminds us that our lives go through so many different cycles of grief, each different and having profound impacts that shape who we are as we move through our lives.

It's not only the athletes that grieve in sport. We see grief attack the fans and supporters as well. England had made it to the 2020 Final of the UEFA European Men's Football Championship. England vs Italy, on England's home turf of Wembley. England began with an impressive goal in the second minute of the match. Italy equalised in the sixty-seventh minute. The match went to a nail-biting round of penalties, and then it was tragically all over for England: Italy had won.

The grief that wildly poured out immediately afterwards resulted in anger, violence, irrational behaviour and verbal attacks. The hostile attack of communal grief, when unmanaged, can very quickly escalate negatively with herd mentality and must be swiftly contained to limit damage and restore order. Perhaps, if we acknowledged this as a grieving process, we would be able to equip those experiencing it to handle it better and more constructively.

## IT'S JUST A MATTER OF TIME

While our lives travel in circles, time is linear and physically travels in one direction – a nice straight line – from the past towards the future. Mentally, however, we have the luxury of being able to time travel. The mind can travel backwards into its deeply complex memory system and forwards into its vivid and wild imagination.

Sometimes, when we get lost in the magic of 'now', we become a silent witness of time standing still. This happens fleetingly when we are fully mindful of the moment, observing reality in its extraordinary state of quiet stillness. Neuroscientists have been studying the effect of mindfulness on the brain and what they have found out has been hugely insightful.

There are two networking systems inside the brain. The extrinsic network is the part of the brain that coordinates active tasks such as doing the weekly shop, making a cup of tea or driving a car. This is the part of the brain that is focused on the 'doing'. Then there is the intrinsic network, also known as the default mode network (DMN).

This is the part of the brain that lights up in an fMRI scanner when the person has completely removed their attention from the outside world and has entered a self-reflective, daydream-like state. This is the part that is focused on 'being'.

Think of it like a pendulum: when the pendulum swings one way, the extrinsic 'doing' part is active, and when it swings the other way, the intrinsic 'being' part is active. Usually human beings are either focused on doing a task, or in daydream mode. We are not usually in both modes at the same time.

However, scientists have found in their research of the DMN that experienced meditators can find a centre point for the pendulum and keep both the 'doing' and 'being' networks active simultaneously.

Zoran Josipovic, a research scientist, moonlighting monk and adjunct professor at New York University, believes this ability to activate both the internal and external networks in the brain at the same time is central to meditators experiencing harmonious feelings of tranquil 'oneness' with their environment. If we take the idea of the still pendulum sitting motionless inside a grandfather clock, this is a great metaphor for how in that moment, for them, time is standing still.

Time doesn't really stand still; the passing of time is a given, and the by-product of the passing of time is ageing. As the clock ticks on, we age. Ageing is scientifically proven to be related to the passing of time.

Interestingly, on the International Space Station it has been found that astronauts age slower because of the time dilation effect, which speaks to – without going into a complicated physics lesson – Einstein's theory of special relativity, which says that time slows down or speeds up depending on how fast you move relative to something else. Essentially, by approaching the speed of light (about 670 million miles per hour) a person inside a spaceship would age slower than their twin at home on Earth. In a sense, this time dilation effect makes astronauts physical time travellers.

For the rest of us, as Earth-bound citizens, we might never get the opportunity to weightlessly float around in space forever as a permanent anti-ageing solution – so we may need to accept that ageing is a fact of life on Earth and come to terms with it being non-negotiable.

Still, I think it's important to view ageing as circular rather than linear – like the growing rings of a tree spreading out and creating a strong trunk of solid width and breadth. Our lives also gather this width and breadth of experience and knowledge that shapes us and makes us who we are today.

Circular ageing is a concept that became apparent to me because of two episodes in my life. In my early twenties, my father lost the ability to care and think for himself as the leukaemia took control of his mind and body and I became his carer for a few months. I became the one who made the big decisions. I became the parent, and he essentially became the child. A few years later, my mother was dying of liver cancer and, at the end, the tables were yet again turned. I lost my parents to the circle of life. In that process, we grieve both the person whose role has changed in our life, and our own loss of role. Caring for someone you love is inextricably bound up with grief, because you start grieving for them before they have gone, and you grieve for yourself and the burden of responsibility you now carry. It's ok to feel like that: it's simply a response to change of identity; once again we are circling around who we were, who we are, who we have become and who we want to be.

For me, now I see life and ageing as not just a circle, but a simple spiral. We grow older and we spiral up through life, and then at some point, what goes up, must come down. It's like an aeroplane taking off and then flying at high altitude, with a bit of turbulence here and there, but most of the time it's a smooth clear flight. At some point, the destination is in sight, and we must all prepare to land. We buckle back in. It could be a nice smooth landing, with a slight bump back on to the tarmac, or a landing that requires quite a bit of sideways banking. Crash landings are also possible. Whatever happens, though, that plane does eventually come back down. The flight is over, the engine switches off and it's time to disembark. Thank you for choosing to fly with life; we hope you had a pleasant flight and a safe onward journey.

In our current Western world, ageing – much like death – has had a lot of bad press: it is viewed as something that we should be preventing, avoiding, escaping and hiding from.

It is as if ageing is a parasitic leech that is around the corner waiting for us to fall into its trap, so it can feed on our healthy functioning cells and suck out all our precious cell juice packed with mitochondria and reduce us to dusty ash.

To protect us from the horrible cell-sucking leech, we are constantly sold the promise of a forever youthful exterior – anti-ageing skin care, whitened teeth, hair removal by lasers, lighter and brighter eyes, lustrous hair, tattooed fake eyebrows, the 'I woke up like this' look with semi-permanent makeup, smoothed-out and contoured cheeks, lifted and tucked wobbly bits, plumped-up pouts, perky bottoms, vajazzled Brazilians, waxed backs and cracks, sucked-out lumps and bumps, hair follicle scalp implants and, of course the old favourite, wrinkle-free statically expressionless faces filled with botulism.

Magazines and billboards around our planet are plastered with digital images of 'perfection', offering the latest 'scientifically proven' products that retouch us in real life: lifting and toning tonics, complexion cleaners, luxurious lotions made from diamond dust and magical potions that work instantly. All this mind- and body-boggling stuff makes up the staggering billion-dollar (£27 billion in 2020 in the UK, to be exact) beauty and cosmetic industry.

We are shown the fashion we buy on digitally doctored 6-foot-plus, size-nothing, teenage models with the abs and pecs of a gladiator and stomachs of steel, and we wonder why the exact same clothes on us don't seem to look like they did on the model!

We can all be tempted by the magic of the photography filter as we apply the 'bit of fun' filters to our faces with the touch of a button and, suddenly, we are ten years younger.

The beauty and fashion industry's *anti-ageing* message is the real trap, and we fall head over heels for it. Just by putting those two words together as acceptable – there lies the problem!

I sound scathing of these industries, I know, but I'm not; I'm just very curious as I observe it from the inside out. Funnily enough, my boyfriend is a high-end fashion and beauty photographic retoucher. He is highly skilled at creating 'beautiful' imagery – so I feel I have some authority to say that, even though it looks like the fashion and beauty industry is beginning to accept ageing, it's not. It's not a genuine acceptance at all.

Unfortunately, the fashion and beauty industry's version of beauty has always been – and will always be – unblemished, slim and *youthful*. Ageing and hairy sagginess is not considered to be beautiful. The only wrinkle genuinely approved of in a high-end fashion magazine is earmarking the page to buy now, and not on the model's face.

The beauty industry is starting to nod begrudgingly at representing ageing by claiming to let us know (in the tiniest disclaimer ever) when the photo has been digitally enhanced. The fashion industry has started to include realistically shaped bodies in their plus-size ranges. Personally, I don't think that this nod is strong enough – it's more of a shrug. The beauty industry has found a way to spread its anti-ageing message to technology too.

## VIRTUAL GRIEF

With a simple swipe or tap in the right app, you can literally look like you did when you were 16. Fun, but slightly freaky too! Also, the same applies the other way around – another simple tap and ta da ... This is you at 86 years old. But this image is hideous, causing you to look away in distaste, feeling your amygdala jump into action and insist that you immediately look for the nearest Botox clinic while downing a jarful of collagen.

In this tech-accelerated world, in which nearly all of us are equipped with professional cameras in our pockets complete with face recognition and fingerprint scanning, it's like we are all hypnotically and happily agreeing to become part cyborg. (Does anyone download, print out and read carefully the full terms and conditions before agreeing to use Google or any social media platforms, or download any apps?) What are we all agreeing to? I'm not a technophobe by the way – I am tech's best user. I use four of the major social-networking sites religiously on a daily basis, I have robots deciding what music I listen to every day and I even have a reminder on my phone to get up and walk around. When the internet wasn't around (yes, I am that old!), I managed to remember to get up and walk around, so I'm not entirely sure why I need reminding all of a sudden ...

I wonder whether, with all this technology at our fingertips and slowly or rapidly (however you look at it) becoming imprinted into our consciousness, we will start grieving our ability to think for ourselves as well as grieving the younger version of ourselves?

What about those very clever video apps that swap out faces in the name of entertainment? Very soon, as these deep-fake images and videos get more and more sophisticated and realistic, we will not be able to tell what is real and what is false. Deep fake is coming to get us and, if we are not careful, we will end up blurring reality and not knowing what to believe. The sayings 'seeing is believing' and 'you have to see it to believe it' are unfortunately crumbling into lies before our very eyes. If the real vs fake debate carries on at the rate it is, I'm sorry to be the bearer of bad news (not fake news): *we will all be grieving the truth.*

What we think of as true or false is the foundation for our belief systems. We believe things that we think to be true. They may not be true for somebody else, and that is why we all have specific and different belief systems – depending on our map of the world and our environmental realities.

Our environmental realities also shape how we think and behave. For example, if we are at work – in a business meeting – our thinking and behaviour would be appropriate for that environment. We wouldn't behave and speak in the same way when we are out with our friends on a Saturday night.

When what we think of as truth changes, this can have a profound effect on our map of the world. Our belief-based realities can start to shift in interesting ways. This can be a positive thing for some people, but a destabilising experience for others.

An example of an immediate belief-based reality shift is in the technological advancement of virtual-reality gaming. Once you place the VR headset on, you are mentally transported into a new world – you lose sense of the real environment you are standing or sitting in and your brain immediately adapts to the alternative reality that is presented to you.

My teenage sons have a VR gaming headset, and it has been interesting to observe the immediate reality and environmental distortion that happens. Our TV/gaming room has a lovely Victorian fireplace with an impressive collection of gin bottle vases with delicate flowers inside them. When the gamer has the headset on, their brain instantly loses all references of the actual room they are in as, in order to quickly orientate into the VR world, they need to begin to disorientate in the real world.

The new VR world has a completely different spatial system, so they could be walking around in a Cambodian jungle fighting off enemy attackers that appear out of nowhere. The gamer's brain, in that intense reality, must quickly adapt to the new 'hostile' environment that is being presented to them and, because they believe in this world, fact and fiction begin to merge. After a while, they can't tell the difference between imagination or reality. Fiction becomes fact.

This blurred-line effect is noticeable when their actions and behaviours don't match their real, physical environment any more – their arms fly up in self-defence as they start leaping around the room, making sudden and huge gestures, and they try to run away from their attacker. You can imagine my face when they first used the headset; for the sake of my lovely vases, we now have a strict room-clearing ritual that must happen before anyone can put that headset on!

With this alternative reality, it can feel very real, even though it is not. This same real vs fantasy idea goes back to fashion and beauty, where a lot of the body images we are presented with are not necessarily real. This distortion of reality can have a very damaging effect on our mental health and exacerbate levels of, what I call, 'comparitis'.

With the blurred line between what is real and what is manipulated or enhanced, the youth of today have started to compare their own bodies and photographs to these unreal images presented to them and, because they will never match up, this unreal expectation of what they should look like begins to eat away at their self-image and self-confidence.

Self-doubt has been a huge problem for clients of my hypnotherapy practice. Sadly, I have been seeing more and more younger kids coming in with body dysmorphia, low self-esteem, damaged self-worth and lack of self-belief. The 'I'm never going to be good enough' belief is responsible for psychological disorders such as self-harm, eating disorders, bulimia and anorexia, and suicidal thoughts.

The youngest suicidal child I have seen in my practice was only 10 years old. It is very scary that in today's world a 10-year-old has become so preoccupied with self-image, and paranoid about how they look and what other people think and say about them, that they are no longer able to enjoy living and believe that ending their existence will be the answer.

Today's boys and girls are all walking around with the internet in their pockets. This can be a positive thing when used well, and can be very dangerous when it starts shaping the way these children think and defines their belief systems in a negative way. It becomes a problem when bad fiction turns into fact and the good facts turn into fiction The following story is fictionalised, but highlights some of the issues that, young or old, we can struggle with when our sense of what is real or not is disbelieved.

Marie's mum got in touch with me explaining how she found a suicide note under her daughter's pillow when she was changing her sheets one morning.

Marie's mum – let's call her Mrs M. – as you can imagine after reading that note, felt as if her mind had exploded into shards of shrapnel while her body froze with fear and, after a few seconds, began violently shaking with adrenaline overload. She was in a state of total shock and utter panic at the same time. Part of her brain was trying to figure out what bit of this was real and what she could do with this information, and the other part was a complete mess, unable to think rationally and logically.

After a few minutes, when she could think straight enough, she found her mobile and immediately called the school. As soon as the receptionist answered, she broke down in floods of tears, relieved to hear another human voice. Luckily for Mrs M., the receptionist who picked up the phone was a Samaritan volunteer and knew how to calm her down enough to be able to find out what was wrong and how she could help. Once the receptionist made sense of Mrs M.'s enquiry, she could help her focus on facts and let go of the fiction.

The facts were that, at this moment, Marie was alive. She was 'fine' – she was in school, in a PE lesson and the receptionist said she could see her from the reception window in the playground running around, happily playing a game of netball. These were all facts.

When our brains are forced to face facts, we move into the logical bit – the part that can see things from a rational point of view. This part of the brain is evidential and less likely to be drawn into an emotional overload.

Once you can talk someone out of their emotional mind, they will find that their own pendulum swings from one side to the other and, after a while, there will be a settling down in the middle. This moment of settling down is called emotional intelligence (EQ): a state of mind in which you can think with logic and compassion at the same time. This state of mind is a highly beneficial state to be in when looking at options, figuring things out and making decisions. It is the state in which the brain can make a proper assessment of a situation and apply a realistic approach to a problem and work on looking for a solution.

The solution that Mrs M. found that day was solution-focused hypnotherapy. It was after the conversation she had with the school that Mrs M. got in touch with me and explained what had happened. I started seeing Marie that week, and saw her for a total of eight weeks, during which she went from being a misunderstood, uncommunicative little girl to a happy-go-lucky and considerate child. She understood the information I was giving to her about her mind, her thoughts and how she could use her brain and imagination in a way that could support her.

The children and adults that end up in my practice all have excellent imaginations. As we have seen, our imaginations can either work for us or against us. The imagination is made up of thoughts, beliefs, experiences, senses, words, pictures, feelings, emotions, memories and dreams. It's a blurry but vivid place in between realities. We can use our imaginations just as vividly in a negative way and, with some training, in a purposefully positive way. When we learn how to use imagination in a positive way, we are able to understand the difference between that which is imagined and that which is real. However, when we get lost inside our imaginations in a negative way, it's like the negative imagination has an additional convincer attached to it, and the danger is we can believe the negative thoughts even when they are not real.

My clients initially arrive at my practice because they have been using their imaginations vividly in a negative way and find themselves trapped in the fight, flight or freeze mode and don't know how to get out. Solution-focused hypnotherapy helps them get into the part of the brain that is rational and reasonable – the left prefrontal cortex (LPfC). This is the part of the brain that lights up in scans of the brains of meditating monks. When a person is engaging the LPfC, this helps the brain to divert the imagination from worst-case scenario and catastrophic thinking to creatively and positively finding solutions and working things out. This shift in imaginative awareness begins their journey into feeling better and increases their wellbeing. The shift into the LPfC also helps their brains release positive neurotransmitters to lift mood (serotonin) and encourage motivation (dopamine).

Marie often used to say to me that what upset her the most was that she was never going to be like 'those girls on Instagram'. Once I explained to her about the concept of Photoshop (which she had never heard of), she began to think about those images differently.

Marie is what my mentor and the author and developer of *Hypnosis Without Trance*, James Tripp, calls a *reality tester*. A reality tester is somebody who likes to see evidence about something before they believe it. I am a reality tester too, so I am always pleased to supply evidence where and when I can. She asked me to show her the photo-editing software. I agreed. I asked her to email me a photo of herself and we would work on that image the next time we saw each other. In our next session, I had my laptop set up with her original image and I photoshopped her into a 'perfectly Insta-worthy' version of herself. While I was doing this her eyes widened and her mouth fell open. She was witnessing her truth being distorted right in front of her eyes. In that session, she had a rapid belief transformation: a seismic shift of her reality as she realised in astonishment that what she believed to be true about those 'perfect' girls had altered. Her world had changed for the better, and after that she was like a different little girl.

Photoshop and other image-editing software are incredible; it is very scary how, when an image is photoshopped well, there is absolutely no way anybody can see the editing or even know what *has* been edited. A skilled photo editor will have mastered the fashion and beauty industry's wishes using advanced processes such as L&D (light and dark) – an astonishing way of very carefully lightening and darkening and manipulating the hue of each pixel on the photograph. This technique cleverly removes bags, blemishes, moles, hairs and smooths out wrinkles and lines without affecting the delicate, intricate pattern of the skin, so that instead of just blurring and softening the image, it stays sharp, and the skin is (as the beauty industry calls it) 'cleaned up'.

There is also an extraordinary tool called liquify, which when used 'artfully' can make someone taller, slim down the waist, round out the bust, chisel the chin, lengthen the thigh or the torso, create the six pack, widen the eyes, fill in the lips and digitally suck out any fatty lumps and bumps without distorting the backgrounds or the patterns of the outfits. It's quite extraordinary to watch!

When I've spoken to my boyfriend about his work, he has told me that when he is retouching a fashion or a beauty image, he no longer sees it as a person or human model; he sees the image as millions of pixels zoomed in on his monitor. Pixels are tiny little square dots of colour and hue – a unit of measurement on a screen which equates to ¹⁄₉₆ of an inch (0.26mm).

When the human models in magazines have been deconstructed and reduced into pixels, this level of microscopic depersonalisation in what we see advertised and displayed around us has surely got to have some impact on what we think about our own image, and therefore how we behave in relation to our own image.

I have no intention of destroying my boyfriend's career here, but it's an interesting conversation to have about how the beauty and fashion industry unknowingly contributes to our potential feelings of subconscious grief for who we used to be, who we thought we were, who we will never be again and who we never have been.

The grief that shows up in this context is quite subtle in some ways but hugely obvious when you think about it. We are constantly surrounded by still and moving images that are not real. They are enhanced, filtered and manipulated. The truth is not only stretched, but also reduced.

We are not used to seeing raw unedited imagery, so if we take a selfie and look at it, there will be a part of our own psyche that is not able to accept the reality of that unfiltered image. The danger is that we then begin to manipulate and edit the truth away ourselves. We can slowly become the retoucher of our own image, and that is a cause for concern.

The names for some of the tools used to enhance images in photo-editing software give us clues as to what they do to the image and how this can then influence our reality and how we see ourselves.

I've mentioned liquify already – the idea of a person turning into a pixel. It has the same reductive quality to it: turning a solid into a liquid – melting the human form down into a digital playground. The warp tool – brilliant for warping the truth and bending reality; then we have the mask tool – great to hide behind and for pretending to be someone else. And then there's the clone tool. Need I say more? The healing brush has a sense of redemption about it, but of course it really means digital healing in a few seconds, not the deep quality of healing that takes time and effort.

If only there was a magic wand (oh, yes, that's another tool) in real life that we could wave over our emotional grief and our real-life scars – wouldn't that be a great world? Hmmm. I think not!

The grief for the past version of ourselves can make us feel inadequate, affect our self-confidence and self-esteem. We may start to dislike what we see in the mirror because the mirror never lies.

I have several female clients who have told me that they don't have mirrors in their bedrooms or bathrooms because they don't want to see what they look like when they are getting undressed.

The fictional case study of Eve is an interesting example of this. Eve, a 30-something dental nurse, came to see me with catoptrophobia – a fear of reflections. She was unable to have any mirrors in her house and even covered up reflective surfaces, such as glass tables, with tablecloths.

In my practice room, there is a small mirror hanging by the door so clients can check their appearance before they leave as when they lie down on the couch their hair can get a bit messed up! This was the first thing Eve saw when I opened the door to her, and she couldn't come into the room until I turned the mirror around. At her third session, I forgot to turn the mirror around, and she walked in and didn't even notice. I knew then things were looking brighter.

It had come to light that her fear had manifested because of her own disgust of her physical appearance. I asked her if she had any idea when this had started, and she nodded and talked me through the time she overheard her school friends talking about her, saying that she was so ugly her face would crack mirrors. She was around 14 years old.

The irony was that, as a dental nurse, she was surrounded by reflective surfaces – the dental surgery had shiny cabinets and glass worktops. The dental apparatus that she had to clean and disinfect in between every patient were mirrored. She managed to overcome her fear at work, but in her private life, this was when it showed up and limited her.

Surprisingly, she was always very well groomed and put together – not sure how she managed to pluck her very neat eyebrows without a mirror, but she did – and her makeup was always subtle, but immaculate.

It only took a few weeks for her self-esteem to grow and self-belief to return and, after a few more weeks, her self-loathing became a thing of the past.

Our behaviour, as we see in this example, can be influenced by things we hear and believe. At the impressionable age of 14, the mind can be a vulnerable place. It can be difficult to move on and let go of thoughts or opinions, and once an opinion gets lodged as a truth, our reality can start to confirm this thought as true.

Once I can help a client understand that opinions are not facts, they can start to let go of those opinions that have caused grief in their lives.

Self-confidence, self-worth, self-belief and self-value are principles that need to be addressed from the inside out. Once a person can discover that they need to learn how to self-validate, they will resist the temptation to look outside themselves for approval, and then whatever happens in the external world – such as other people's opinions and ageing itself – the person will always be able to separate fact from fiction.

It makes me sad to think that body dysmorphic states of mind arise from a person's inability to move on with how they think or to let go of what was, and from their desire to change something that they have no real control over, like ageing.

I feel very strongly about re-educating the planet to view ageing as a positive experience – one that we can learn from and treat with reverence and respect. Here's an idea – what if we came to understand the ageing process from a scientific perspective to see if there is anything to be done to prevent our cells from ageing, rather than throwing money and resistance at it when it's already happened?

Dr David Sinclair is a professor in the Department of Genetics at Harvard Medical School and he is best known for his work on understanding why we age and how to slow down its effect. He tells Joe Rogan on YouTube that he has found a simple (it sounds complicated to me) formula to activate the body's defences to ageing. If you are ready to hear a potted version of his research, I suggest getting your chemistry head screwed on and concentrate while you read the next few paragraphs.

In his extensive research, Dr Sinclair found there are genes called Sirtuin, which he makes a point of saying are *not* anti-ageing genes, but longevity genes.

We have seven of these Sirtuin genes, whose function is to protect all organisms on the planet – from bacteria to plant life to human beings – from deterioration and disease. The Sirtuin gene senses when we are hungry and when we are exercising, and knows when to send out the cellular troops to defend us. Basically, it keeps us in balance.

Dr Sinclair found that when you add more of these Sirtuin genes into yeast or a mouse, they live between 5 and 20 per cent longer! He then went on to test how these genes work with human molecules. In these tests, he found NMN (Nicotinamide Mononucleotide) and Resveratrol are the molecules in the body which also mimic and contribute to the longevity effect. Resveratrol is the accelerator pedal for the Sirtuin gene, and the NMN and another molecule called NR (Nicotinamide Riboside) are the fuel. Are you still with me? Don't worry if you're not. It's complicated.

The Sirtuin genes also need another very important molecule called NAD (Nicotinamide Adenine Dinucleotide), which is central for the metabolism to work. Without NAD we would all be dead in roughly thirty seconds. As we get older, we lose NAD and, by the time we are 50 years old, we have half the levels of NAD that we had when we were 20. Essentially, our metabolisms slow down as we get older. Dr Sinclair found that NMN and NR boost our levels of NAD back up again. When this was tested on mice, it was shown that with increased levels of NMN and NR, they lived longer and were super healthy.

Dr Sinclair says that all life is governed by the same universal regulators of ageing. There are three main pathways that respond to how we eat and how we exercise.

The first pathway is called AMPK, which is targeted and boosted by Metformin (a drug that is prescribed to diabetics, although not to combat ageing), which increases longevity. AMPK's job is to send out the troops to prevent disease.

The second pathway is the Sirtuin gene that keeps us in balance and, remember, it needs our NAD (metabolism) to work, and the NMN and Resveratrol molecules help boost that.

The third pathway is mTOR – this is the growth and survival gene. This pathway checks and responds with amino acids that are in the body, which also protect the body with the right tweaks.

So, if you understand all of that – well done, because I didn't.

All I know is that our body seems to be a very clever chemistry set, and when you are a very clever chemical scientist, you can change things to help in a positive way!

Another longevity tip that Dr Sinclair has for us is to give the body some temperature stress. The science around this has led to an interest in ice baths and cold showers among some fitness enthusiasts, but should be treated with caution unless you know what you're doing. However, this practice increases the lifespan of yeast: in the lab, when subjected to extreme heat and extreme cold, it has been found that yeast cells live 30 per cent longer.

After listening to hours of lectures on this topic, it seems sensible to put our energy into helping our bodies and minds stay young rather than apply a deep-fake filter, because a real-life deep-fake filter does not exist!

There is also power in people. If we as people understand that ageing is a given, unless we all start our own longevity experiments on ourselves (don't try this at home, kids!), we can begin to make different consumer choices about which 'model' we subscribe to – the retouched one, the experimental one or the one of acceptance.

If we can learn to grow old gracefully and graciously, we will save energy, money and time – we will finally stop trying to fight a system that will always ultimately win.

We will be able to look at ourselves closely and accept ourselves no matter what is being reflected at us in the mirror. We will not continue to lose heart when another grey hair pops out with enthusiasm for life or another wrinkle waves a happy hello; we will embrace our ever-changing bodies and learn to love them because they will continue to carry us around until we don't wake up any more.

Our faces are the screen of our lives, lived. We don't have to keep our faces stuck inside the screens. Our wrinkles are all the times we have smiled, and our genuine smiles will return once again. The lines on our foreheads are the grooves for our incredible emotions to live in. Our emotions are the lines of communication that connect us to all the human beings on Earth, and as human beings we can come together in this appreciation of living longer on our planet that also ages beautifully.

Nature is full of ageing cycles. Think of the delicate little apple-white flowers blossoming into the fruit that falls off the tree, eventually turning to compost to nourish the soil again, and to be the host of new growth and new life.

Where there is growth, there is always death, and ageing I guess reminds us of that inevitability. We are reminded of our own mortality here on Earth, and that, to some of us, can be uncomfortable.

We can grieve our short time here and worry about how much time we have left, or we can embrace each day we have and be grateful for those we have already had and to wake up to another one. Another day older means another day we get to spend on planet Earth, and we become another day wiser, turning the history lessons of yesterday into the informed choices of tomorrow.

Ageing is what makes us human. We are not robots, avatars or pixels on a screen. We are amazing natural creatures who have the capacity to love and accept who we are and everything that comes with that.

Sometimes, you just have to pick your battles and the battle of ageing is an age-old one; unless you are going to become an astronaut and live on another planet out of our current orbit for the rest of your days, I'm afraid ageing is something that you will have to make peace with, and you may find that the more you accept ageing, the younger you will look and feel – without spending a single penny and without having to start life again on Mars.

You'll be pleased to hear it's not all doom and gloom for the silver fox – I have found some pro-ageing cultures! There are a few places on the planet where ageing is culturally viewed as positive and in some cases welcomed with open arms.

In India and China, elders are viewed with respect and given authority as the head of the family. The older you get, the more wisdom you can impart, and you are often seen as the peacemaker, problem-solver, matchmaker and general fount of all knowledge.

Here, the elderly are never discarded into a care home or rendered useless; in fact, it is considered sinful to abandon your parents or those older than you. In these two cultures, as well as in Korean tradition, the elders are a huge asset to the family – often providing childcare for their grandchildren as their own sons and daughters work to support the family. Much of the Korean regard for ageing is rooted in the Confucian principle of filial piety – a fundamental value dictating that one must respect one's parents.

The extended family system does seem to have benefits for the whole family unit. When you look at it logically, it makes sense from an ecological, financial, emotional and socio-economic perspective, but it has been ingrained into our Western culture to live independently and fly from the parental nest as soon as we can, and that seems to be the measure of our success and achievement.

Interestingly, though, in Britain, according to a 2019 article by James Tapper in *The Guardian*, it is thought that there has been a rise in families bunking in with each other. In 2019, there were 1.8 million families who chose to live together in multi-generational households.

This sounds like a positive move but, alas, it's not because young British people are hearing about Confucian principles and suddenly changing their values by caring for their ageing parents; it's because young people cannot afford their own homes and so the bank of mum and dad becomes appealing.

There are mixed feelings about this – some people say that the extended family living has caused them to feel more depressed, but then there are other stories where people have reported reduced feelings of isolation and loneliness, leading to increased wellbeing and overall happiness.

This perhaps demonstrates that for multi-generational house-holds to work well, there must be a cultural mindset shift, in which each person is assigned a role and the family dynamics must be clear and readdressed. Clearly, it must come back to learning and applying some tribal basics.

Many tribal communities in Africa and indigenous populations, such as Native Americans, also have the same respect and admiration for the elderly. Often the elderly continue to contribute to the running of the tribe well into their old age and are often the most revered members as they pass on the traditions to keep the tribe alive. They have tribal court hearings around a fire to hear the disputes and when the elder of the tribe passes judgement that is the final word, and everybody respects that decision.

Coming back, closer to the West, the cultural stigmas around ageing and death don't seem to exist in Greece. In Greek and Greek–American culture, old age is honoured and celebrated, and, very much like in India, China, Korea and the African tribes, respect for elders is central to the family unit.

I was very close to my grandmother when I was growing up, and I used to go and stay with her regularly. I called her Diddima: Bengali for grandparent on the maternal side.

I loved staying at Diddima's because we would always sleep together in her bed that smelled of sandalwood and musk, and she would tell me so many interesting stories as I fell asleep.

One morning, when I woke up, still with a smile on my face from the memory of the heartwarming stories told the night before, I asked to see her story book. She looked confused and asked me which story book I meant. 'The story book that you get all the stories you tell me from ...' She laughed out loud – her laugh was infectious, and I can even hear it now as I type these words: a kind of cheeky, wise, old, joyful cackle. I laughed along too, although I wasn't sure why my innocent request had tickled her so much. She replied, 'Beti [loving term for child in Hindi], I don't have a story book you can hold. All the stories I tell you live inside my heart and each of my heartbeats are the pages of my story book.' I must have been about 6 years old, and I remember thinking, *Wow, I can't wait till I have collected enough heartbeats to have that many stories living inside me.*

Storytelling is what connects us as a species. First through our ability to think up the stories; second through our ability to tell them; and third through our capacity to listen to them and feel like we are in them as we hear them come alive in our minds.

I'm thinking about how powerful stories have been important to my own development and function. From when I was a small child, the wild and wonderful stories told by my grandmother captivated my imagination and took me on journeys across the world, into the deserts and over the seas, away to secret faraway lands and exotic places that were coloured and populated by my own mind. I would often go off to these fabulous places and my brain would be there to simulate a full-body experience. Even though I never learnt to swim, I was able to dive into the oceans with the mermaids and explore their magical castles deep underwater and then, if I wanted to fly like an eagle, I would soar high above the treetops. I would meet strange-looking mythical creatures and incredible magical beings. I would have no sense of time and space – it was like a million worlds opened up all at once. Everything was always so colourful and bright. The colours seemed to pop and sparkle. The worlds would be peppered with possibility, curiosity and wonder. It was such an important part of who I was.

This passing down of creativity shaped me and allowed me to let go of the sense that I was a fixed being with only one reality.

# ... AND LETTING GO

In an anti-ageing, pixelated world, it's no wonder that we all subconsciously have a fear of growing old, of letting our greys shine through, and feel anxious when we find another wrinkle to add to the collection of smiles. We stop smiling. We stop loving. We start to grieve our past. We crave our youth; we long for the lean and streamlined bodies we took for granted back then and wish for our teenage metabolism again. We really struggle to find ways of letting go of these idealised versions of ourselves and to move forward with our lives.

Talking of teenagers – if you are not a teenager reading this – can you remember what you were like back then? Specifically, how you viewed the past and the future ... Just take a moment to think back and see if you can remember.

I don't remember being overly concerned or worried about my future as a teenager, but I do remember vividly fantasising about it – who I would become and where in the world I would live. I imagined becoming a famous ballerina, even though I only went to one week of classes. I watched *Fame* and wanted to be just like them.

I also remember being rather attached to my past. I found it difficult to let go of old friendships that had fallen by the wayside and spent a lot of my time writing about stuff that I had done to document it somehow in a historical journal. I was a serial diary writer. I loved writing, so I did it all the time.

I had twelve pen-pals on the go at one time. I would write letters to all sorts of people from all over the world. There was a magazine called *Just Seventeen* that had a pen-pal section in the back, and I used to get this magazine at 14 and felt so grown up – especially when I wrote off to all these PO boxes and then, to my utter surprise and delight, got equally enthusiastic letters back, from real people, from faraway lands. There was Andrea in Australia, Petra in Germany, Kasha in Arizona, Frank in Switzerland, Emile in Paris, Raju in Essex, Debbie in California, Patrick in Golders Green, Michael in Canada, Aarti in Bombay, Phoebe in Scotland and Anya in Poland.

I used to look forward to the post every morning, because invariably there would be at least one letter from someone to pore over. I got lost in their worlds, their words. I imagined my life as their life. I lived everything they were telling me – about their schools, their hobbies, their pets, their siblings and their travels.

My imagination was wild, and I was being filled with adventure and possibility. The world was on my doormat and I spent a lot of the time in my own little Dipti bubble – off in a daydream somewhere or chatting incessantly about all these different people. I loved to tell stories. I loved to dream, and I loved to talk about my dreams.

As I think back, I remember in every parents' evening the teachers would say something along these lines to my mum and dad: 'If Dipti just stopped talking so much, she could be very bright – she won't get anywhere in life if she keeps looking out of the window or talking so much.' Then my mum would make the comment about me being chucked out of the ballet class at 3 years old for being disruptive, and how some things will never change – 'there's no hope for that girl ...' My dad would sit there, mortified, and almost hang his head in apologetic shame – and then dart me a look to suggest that there would definitely be 'words' tonight Missy.

Isn't it interesting that I am now a successful hypnotherapist, relaxation teacher and lecturer, and am invited to speak at major conferences and to be a talking head on debate programmes? Talking and dreaming has become my main source of income and has got me to some very interesting places in life!

However, my brain in those painful parent–teacher appointments was not concocting a plan of what my future career was going to be, and I wasn't informed and eloquent enough to speak up for myself and explain that talking would get me far in life and that the brightness of my future might be governed by how much I talk and dream, but not in the way they were suggesting. Instead, I just used to sit there, mute. For once, I stopped talking. I was taught to be silent when with people in 'authority' – for me, this meant teachers, people in uniform, judges, bosses, doctors, managers and anybody legal or in law enforcement.

How had I come to define these people as 'authorities'? Well, as children we all learn subconsciously by how our parents act and behave. We also learn from what they say, but behaviour is often what we first reference, as it is an inbuilt function of the brain's mirror neurone system.

My dad started having heart attacks from when I was 3 years old. His first one almost cost his life but, because of the carefully balanced cocktail of drugs that he was on, he just miraculously kept going.

I was quite young when I used to go to visit the doctor with my dad. I remember there were many visits, often one every week. The doctor's surgery was musty and dark – it had this strange lingering odour; I never knew what it was until I recently had surgery myself and my dressing smelled of it afterwards. I asked the nurse what the smell was, and she said it was iodine.

The smell instantly transported me back into that doctor's surgery. There was a massive mahogany desk, with a chocolate-brown leather tabletop secured down on each side with these giant drawing pins. I used to run my fingers along the smooth and bumpy surface; feeling the cool metal with each fingertip was quite nice in the hot, stuffy room. The doctor was an ancient-looking Indian gentleman, and he was very, very serious. I don't think I ever saw him smile. He was slightly deaf and needed things to be repeated two or three times.

It was the same drill every time. Before the visit, I was told numerous times to stay quiet and be good. When the buzzer sounded for our turn, my dad would stand up, clear his throat and then, as soon as he crossed over the threshold from the waiting room into the stuffy, iodine-smelling office, he acted as if he had entered the celestial parlour of an omnipotent being. He would wait to ask to be seated. (I never sat because there was only one chair so I stood to the side, and that's how I know exactly how many drawing pins there were on the side of the table – twenty-three, although there should have been twenty-four but one was missing.) In a small voice – that my dad never used with me – he would explain his symptoms to the doctor and would have to repeat himself many times as the doctor couldn't hear a thing. Dad would always call him 'Sir', profusely apologise and stutter his way through the appointment – which was probably only a few minutes long, but it honestly felt like hours to me. It was a rather painful process to be part of. Dad would always almost bow down to him in gratitude after he had finished speaking.

Then the perfunctory checks. Blood pressure with the black hand pump that looked like the hooter bell on my bicycle and the stethoscope that he used to rub vigorously on the palm of his hand to warm it up before he used it on my dad's chest. And then the beeping machine which needed a lot of blowing into. Not sure what that one was for!

After all the checks were complete, the doctor would mumble in a very confusing fashion as he reached for his pad of special magical paper. He had an ink-pen that he would dip into a pot of indigo ink and draw these squiggles in magnificently flowing script all over the paper. He would then dramatically and artfully sign the paper and offer it to me like a hanky, with the instruction to wave it around until it dried. That was the bit I looked forward to, waving that magic bit of paper around. Then Dad and I would walk back to the car park down a cobbled pathway, and I remember jumping from one cobblestone to another feeling grateful for the piece of paper in my hand, because that was the magic paper that we would trade in for a couple of shaker jars that made a nice sound. Like maracas. Those magic pills and shaky bottles kept my dad alive.

I learned from a very early age that the doctor is someone to be taken very seriously – he is a very important man in life and everything he says is gospel because he is a life-saver and life-lengthener. He is like a god – a god that smells of iodine.

My dad used to say to me every day he and I wake up is another day we are both still alive and that is something to be thankful for. Death was a daily conversation in my world. Not a morbid conversation but one that was often referenced and pointed to. The concept of dying was normalised and discussed in a factual way. I don't remember being worried about death or anyone dying, but I do remember thinking that everyone must think about death in the same way as our family did.

I was always reminded of the story of my own unusual birth from an early age, and I think the 'you nearly died as a baby' story is always going to be something that stays with you for your whole life.

I was born prematurely, at thirty weeks – that's ten weeks early – and I weighed in at 2 pounds, 4 ounces. This was in 1973 and, according to the 'Born too Soon' report by Jane Griffin, senior research associate for the Office of Health Economics, pre-term births around that time had a 50 per cent survival rate and 10 per cent of the pre-term babies that survived had lung disorders. I was one of the survivors, and one of the 10 per cent with severely underdeveloped lung function.

Not a terrible trade off, but it's been a reminder of how important it is to be able to breathe. Breathing is something that most of us take for granted, but when you have not been able to breathe as much as I have, you learn to let go of the fear of death. I have developed this unusual skill to relax enough during an acute asthma attack and imagine I am still breathing with my mind.

I've had quite a few close calls with death myself and have learnt that, each time, it gives me another interesting story to tell. (That's why I don't go to parties much – my stories are the kind that sober you up!)

The same year my dad had his first heart attack, I had my first asthma attack. That year, both of us found ourselves in intensive care, both learning that death is possible.

The series of his heart attacks and my asthma attacks continued like this for many years.

My mum was also not a well person, but she didn't have a freedom pass into the hospital like me and Dad. She had her own health problems that were formally undiagnosed, but now, looking back with a therapist's perspective, she had her own myriad psychological issues and physical deformities, as well as some serious limiting belief disorders.

My parents were not so keen to talk about mental health; this was a subject that was very much brushed under the carpet. It's almost as if mental health issues were something to be ashamed of and disregarded.

I used to watch a programme called *Tomorrow's World*; it was the 1980s, and technology was starting to take off with very basic computer-generated imagery. This programme helped me develop a very keen interest in scientific brain health. Like the words from my pen-pals, I swallowed up all this information and became fascinated by neuroscience, although, at that time, when I was in my early teens, I didn't know that was what it was called. I learnt about stress-related anxiety from Judith Hann's description of it on *Tomorrow's World* one day and this explanation, accompanied by the computer-generated model of the brain and its functionality, somehow stuck.

Some years later, when I was halfway through my A-levels, I felt myself getting very anxious and incredibly stressed because I absolutely hated two of the subjects I was doing.

My dad had persuaded me to take economics and history because he thought that they would be good for me. I insisted that I also do English, because of my love of reading and writing, and thankfully he agreed, and said I could have one choice!

About six months in, I started struggling. I couldn't cope with two-thirds of my workload because I had no interest in the two subjects my dad had suggested I take, and all I wanted to do was stop and start again the following year with different subjects. I wasn't allowed to do this, and this caused so much tension and many arguments in our home.

My tears were frowned upon. Crying was not allowed. So, I discovered a clever system of swallowing back my tears rather than letting them fall out of my eyes. In Panjabi, there is a saying Rōnā – which means 'don't cry' – I used to hear this a lot.

I didn't cry openly but inside my tears felt like they were drowning all my positivity. My mind had slowly turned into a negative place. All the bright colours began to get duller and I hated my life, my parents and myself. I remembered Judith's words about stress-related anxiety, and I recognised all the stress symptoms in myself. My parents didn't want to hear the word stress come out of my mouth, and they both told me that 'stress' was simply a good excuse for me not to do any work. Stress was not *real* (according to my parents), it was a modern-day justification of laziness. With my head in my hands, and clearly getting no sympathy or useful help from my parents, I spoke to a teacher, who advised me to contact a medical professional. So, I booked myself in to see the iodine-infused Doctor God.

It had been a few years since I had last been there, and I got to sit in his patient chair this time. I told him about my anxious thoughts and feelings of stress and asked him if he could help me be more positive. I told him that I didn't want to do my A-levels any more because I wasn't enjoying them, and they were stressing me out because I felt academic pressure to do well with subjects I hadn't chosen myself. He just looked at me with a confused expression and said that I am far too young to be stressed and I need to stop being hysterical and continue to focus on my studies. He finished off by getting his magic pad out and writing me a prescription for valium (which I had heard my grandma talking about as being the medicine for those people who have a 'broken brain'). I sat there, speechless. A memory came back to me. I had totally forgotten that Doctor God didn't deal with the mind and only seemed comfortable dealing with matters of the body.

I remembered back to a time when my mum took me to see him when I was 7 and still wetting the bed – he said to her that my bed wetting was not stress related, but rather because (like my lungs) my bladder was weak and had been late in developing. Then he gave me a magic prescription to wave around for a packet of incontinence bed pads.

This time, walking out of his surgery with my 'broken brain' prescription, I felt desperately alone and knew that my brain wasn't *broken* – it just needed to find another solution. So, before the ink even had time to dry, I did something different with that bit of paper. I tore his ink-penned script up into tiny little pieces and put all the bits into the bin and decided never to go back to him again. I felt totally isolated and lost, but I was determined to find a solution, and at that desperate point, I didn't care what the solution was. I needed a way out of my situation, and fast.

I went home and sobbed. I sobbed for hours on my bed, my silent tears being muffled by my pillow. I cried myself to sleep and dreamt about being suffocated with my own tear-drenched pillow. My dream was so vivid and clear. I couldn't breathe, I couldn't shout out. No one could hear me. I was drowning in my own ocean of tears and being supressed by my own pillow. It was a horrible, horrible night. I prayed that night for a miracle.

Miracles happen in the most unexpected ways.

The next day was a Saturday. I heard my mum come into my room to bring me a cup of coffee. For some reason, I couldn't open my eyes. I felt awake, but my eyes stayed shut. She was calling my name and I could hear her voice getting closer to me, and I just felt unable to move or wake up. It felt like a state of sleep paralysis, which I sometimes got in the middle of the night. The next thing that happened made me realise I was not paralysed. She screamed and I heard a crash, and I felt a hot splash on my arm and the side of my face. She must have dropped my cup of coffee on the floor and ran out to get my dad. My dad came in, then they were both freaking out. There was talk of calling an ambulance. I also started freaking out in my head, because I couldn't move, couldn't talk and still couldn't open my eyes.

I was most definitely awake, because I could hear everything and still feel the drip of liquid travelling down my arm and down the side of my neck. My brain was scanning all the possibilities of what could have happened. Stroke? Coma? Lucid nightmare? I heard them calling 999, and it was then I heard them tell the operator what was going on. I had a leaking body and a raw crusty face. Like a lizard. I was unconscious (I wasn't), and they didn't know if I was alive. The ambulance was there within ten minutes.

I was taken into Central London, to Westminster Hospital, because they specialised in dermatology. It turned out that, overnight, I had developed (out of nowhere, I might add) a severe atopic eczema that had broken through from the top of my head down to the tips of my toes. The eczema had wept out in the night and dried, sticking me to my bed, rendering me immobile. My eyes and mouth were literally glued shut and my face was crusted over.

I was a medical anomaly and on my paperwork, underneath my name, it actually described me as a medical alien. The consultant dermatologist at the Daniel Turner Clinic at Westminster Hospital immediately inducted me into a medical trial, as they had never seen anything quite like this, and I became a case study in one of the leading dermatology journals.

I remained in hospital for just over two months. I was a mummy, wrapped in steroid-soaked bandages every few hours, drip fed immunosuppressants and trained in their innovative programme of clicker-controlled itch resistance. It was horrible, but I much preferred this to my A-levels.

This is the weird miracle part ... it just so happened that missing two months of your A-levels means you forfeit the year and must begin it again.

While I was lying mummified in that hospital bed, I realised three very significant things that would change the way I thought about my body and my mind forever. Firstly, I realised that my emotions were a signalling system for my body to pay attention to. I then realised that, through these signals, my body was helping me by giving me more obvious symptoms for me and others to pay attention to. Lastly, I realised that with all of this going on my mind was learning how to see the symptoms as messages. The message, I learnt, was loud and clear. The mind, when given the space to think laterally, has always got the solutions. Ultimately, I realised that this interplay between our emotional and physical beings is the most perfect collaboration.

After that most peculiar incident, I started my A-levels again, and swapped out history for sociology, and wanted to swap out economics for biology, but because of a timetable issue I couldn't so I stuck with economics. It was a difficult year, but I got through it, and that was when my avid interest in alternative and complementary healthcare began.

I knew that my brain and my body were communicating loud and clear that Saturday morning: my body was finally crying the tears that my eyes were not allowed to.

My dad had another major heart attack while I was doing my A-levels. It was a close call. When he came out of hospital, he said to me, 'You know, I was lucky again. One day, I won't come home.' I didn't want to believe him, and, when he was right, it threw my world into turmoil. One day, my dad simply never returned home.

Even though death had been part of our conversations since I was small, I also had the evidence of my dad having heart attack after heart attack and then always coming out of hospital – apparently 'right as rain' – so, somewhere in my psyche, I believed that he would always bounce back and he must be immortal.

When he died, I was in my early twenties and the first emotion that hit me like a bullet in my belly was anger. I was angry that he didn't listen to me about stress. I was angry that he left me without saying goodbye. I was angry that he didn't take me seriously. I was angry that he worshipped Doctor God. I was angry that no one taught me about grief. I was angry at myself for being angry. I was

angry with my mum for not learning how to drive. I was angry with
the planet for not prioritising relaxation.

Anger is a primitive way of increasing our strength. When har-
nessed and channelled in a productive way, anger can be the rocket
fuel that helps us speak up, change our ways and move mountains.

Anger, like the lava from a volcanic eruption, violently clears and
reshapes the landscape; it burns through forests, melts blocks and
obstacles, and solidifies in the oceans and creates new islands. Once
the intensity of anger dies down, we can begin to explore and discover new territory without fear or anxiety stopping us.

Once my anger subsided, I realised I was right and had been right
all along – that there *was* this invisible thing in life called stress, and
that stress was a silent and deadly killer. I realised that stress was
a taboo subject in my family and talking about mental health was
even more taboo, not just in my family, as it turns out, but notori-
ously in the Indian community.

There is no Hindi word for depression. When I looked it up on
Google Translate, the Hindi word that came up was *dipreshan*, which
is basically the English word with an accent added. I couldn't believe
that I'd never heard this word before. It is not openly acknowledged
or discussed in the Indian community, and there is a lot of shame
and stigma surrounding not only *dipreshan* but also anxiety.

There is a word for anxiety, which is *chinta*, so it seems as if anxiety
is acknowledged, but *dipreshan* isn't. There is also a word for stress,
which is *tanaav*. It's rather interesting to perhaps consider why the
Indian culture may seem marginally more accepting of stress and
anxiety as a concept, but not *dipreshan*. My theory is that it may lie
in the Hindu belief of suicide being a sinful act, which is true of many
other cultures and religions.

In Hinduism, suicide is simply spiritually unacceptable. Generally,
taking your own life is considered a violation of the code of *ahimsa*
(non-violence) and is considered equally as sinful as murdering
another person. Perhaps there is a deep-seated fear of depression
because it is seen as the slippery slope down into the irrational mind,

and, for Hindus, the irrational mind is a dark place that contains the very sinful suicidal thoughts. The easiest thing to do is turn a blind eye, pretend it doesn't exist and hope that it will go away.

Depression doesn't go away. And I know this, and that is why from that anger came a desire to explore the idea of therapy and grief and start talking about grief to everyone I could.

My family were used to me being different, and they watched from a healthy distance as my world started to change as I began to carve a very new pathway for myself. A pathway that led to my first book, *Good Grief*, and my commitment to mental health and hypnotherapy.

I'm still not sure if they understand what I do – because I'm not a real medic, I'm not a doctor. But they do know I seem to be successful, and that's all that matters to them!

I was always the black sheep – not afraid of having conversations about taboo subjects, I continued to delve deeper into the psyche of the human mind. What I have found there has made for an interesting journey, that's for sure.

Stress is a catalyst, convincer and a motivator. It arrives into our awareness as a mental state of mind; it then gets broken down into a series of thoughts that either convert into anxiety or fear, and hold us back or drive us forward into taking action.

If stress thoughts are not dealt with appropriately and convert into unresolved anxiety and fear, they will eventually leak into the system of the body and show up in various diseases, disorders and physical symptoms. I have seen this first-hand with what happened to both my parents and also myself.

I was 19 when I explained my stress theory to my dad one day. I also said that he should change his GP because his didn't deal with stress-related issues properly and kept giving him tablets to keep his heart beating but no real advice for preventing further attacks.

He looked at me as if I had grown two heads. He told me that I didn't know what I was talking about and that Doctor God had over fifty years in practice and did I know how long it takes to study medicine and that the doctors know everything and if stress was a cause for concern he would have mentioned it. That conversation wasn't great, because it ended up with both of us getting stressed and shouting at each other: two mindsets fixed on opposing ideas.

My words didn't save my dad's life, or my mum's. So, I have had to make peace with my past and turn my present and future into helping other people, and that was the catalyst for me becoming a therapist – and, of course, it had to be a talk therapist!

I learnt about stress the hard way – by being stressed. Stress put me, my mum and my dad into hospital a lot. Stress and those flashing blue lights and wailing sirens were bedfellows.

My dad's heart attacks coincided with times of stress, but no one seemed to see this pattern apart from me, and I couldn't do anything about it then: I didn't have a loud enough voice, my words were never heard and when I tried to speak, I was ridiculed. But I believe to this day that stress took my dad away from me, and that is why I have learnt the neuroscience about how stress affects the brain and then how this effect can show up in the body – like the day my body cried all seventeen years' worth of my stored-up tears. It's a clever system. And I now teach this system to all my clients.

The process of letting go of stress can take us out of our grief state into one of acceptance, in which we can embrace moving forward.

## JO'S STORY

Jo's first experience of feeling grief came, aged 15, when her family dog got run over – she describes her feelings as an unexpected raw, intense pain, and she tried to explain to her school friends about her pain but, at that age, her friends didn't understand.

She recalled this same feeling as her grandma passed away. She was given an opportunity to say goodbye to her grandma, but she declined, believing that she wasn't going to die. Sadly, she did die, and Jo never got to say goodbye, which she still deeply regrets. Jo explains that, again, this sense of raw, intense pain with a big dollop of guilt flooded her system. She now recognised this feeling as grief and was able to use this feeling of deep sorrow and regret in a positive way.

Her mum was dying in New Zealand and at that time Jo lived on the other side of the world in London. She had just come back from a wellbeing retreat in France when she got the news that her mum was in her final days. So, Jo booked her flight and she had her fingers and toes crossed that her mum would be alive when she arrived. Her mum was in a coma and Jo knew that she could still hear. Jo instructed her sisters to keep telling their mother that Jo was on her way and to wait for her.

She did wait for her. Jo was able to spend twelve hours with her mum, taking turns to 'do shifts' that night with all her sisters in the room. Jo went to have a shower in the morning and no one else was in the room; it was at that moment that their mother took her last breath. It's as if the dying person knows exactly the right conditions to make their exit. Which is sometimes a reassuring thought. Death can be, for some people, a very private affair. When we are born, we may not get a choice of who is in the room, but when we die, it's nice to know that we may have a choice – it can be a very special time and to be included in a final goodbye is such a beautiful honour. Death can also be a very bonding and uniting experience.

Before her mother died, Jo's niece picked her up from the airport when she landed in New Zealand and they headed straight to the hospital. Jo asked her niece if she was going to go in to say goodbye, and her niece said she didn't want to because she was scared and uncomfortable. Jo told her the story about when she had the chance to say goodbye to her own grandmother and had declined, and how she faced an enormous sense of regret afterwards and that, if she could turn the clock back and say goodbye to her grandma, she would. In the end, her niece went in with her to say goodbye. Later, Jo asked her niece if she was glad she did, and her niece said she was very grateful that she had – I think this is such a lovely story.

When you are young, you haven't learnt how to process grief yet. Jo describes these 'normal' experiences that we go through in the modern world as stored grief: failing exams, pets dying, moving schools, losing friendships, rejections. She explains that the stored grief sits in the system until the 'biggie' comes along and then – BAM! The grief can bowl you over.

For Jo, her biggie was losing her first baby at 24 weeks. This was the catalyst that knocked her sideways. She was working in the corporate world at the time and took three weeks off after her miscarriage. When she returned to work, she describes being 'not herself' any more. She was a shadow of her former self. She didn't have enthusiasm and drive for her job any more, and she realised that the corporate world was not her world. She had changed. The CEO came in one day telling them about his big dreams for the company, and it was then she had the powerful realisation that his dream was not her dream. She had this profound 'a-ha!' moment and left the corporate world. She retrained as an aromatherapist and started her journey as a therapist. She is now an emotional loss specialist.

Her grief – particularly the tragic loss of her baby – helped her realise that life is too short not to follow her heart. Facing death and loss up close shifted her priorities and, at 36 years old, she realised that she wanted to make a difference in the world. The loss of her baby made her change lanes in her life: she shifted from the middle lane to the fast lane and began to explore self-development.

She describes the urgency of feeling that life was passing her by and that she had to make significant changes in her life choices to give her a greater sense of meaning and purpose.

Loss has tremendous power and forces us to make changes – it's what you do with that change that can make the big difference to how your life plays out. Some people use the change productively and fuel themselves to move in a different direction, and others can feel stuck – almost paralysed in fear of the change.

For Jo, grief and loss were a huge catalyst for movement, urgency and change, and what she went through with her own stored grief and loss of her baby forced her to go inward. She changed her mind about spirituality being 'New Age hippy stuff' and has now embraced a philosophical spiritual understanding of loss and grief.

She began to explore Buddhist principles of loss and grief. She describes speaking to a friend who practised Buddhism and told her about the Buddhist (and Hindu) belief that every baby, even in the womb, has a soul. Each soul comes into this life to do the work it's meant to do, and sometimes it needs 97 years to do the work, and sometimes it only need a few weeks to do its work. Her friend said to her that maybe her baby's soul was here to help her change direction in life and bring her a valuable lesson.

This teaching was a huge comfort to Jo. She had never been exposed to much Eastern philosophy, but this Buddhist teaching ultimately helped her process her loss and gave her some sense of meaning and reassurance.

From our conversation, Jo's last words about grief are, 'You've got to go into it and embrace grief, otherwise it will come out of you in a way that may not be useful. Tackle it head on and you will come through it, although you don't get over it. You become it and you come through the other side. Grief is like a big ball inside a box. The sides of the box are pain points. The ball is big at the beginning, and is always touching the sides. But, with time, the ball shrinks and the pain is not so obvious, but then sometimes the ball grows again and the pain points are activated.'

For Jo, the solution is to accept that pain and discomfort are part of life and, if grief is holding and weighing you down, there are some lovely techniques that she suggests trying.

The first is to imagine that the grief load is being lightened. The second is to separate the emotions in an extraction process – by doing so, you give yourself permission to begin the process of letting go of the stuff that is weighing you down. By letting go of the heavier emotional baggage you begin to feel in control, but you also keep hold of the connection, the fond memories and the love.

In this Jo is right: as much as we grieve loss and change, we can reframe that loss to become the catalyst for positive transformation, and also remember that the things that really matter to us in our relationships – not only with others but also with ourselves – are never really lost.

Often in grief, we torture ourselves with the concept of 'letting go' as meaning that we don't care and that 'moving on' takes us away from that which we have loved and somehow lessens the love. These thoughts come from a place of fear – fear that if we stop being over-whelmed by grief, all that is left to us is emptiness. That is far from the truth: we are not letting go of the love we have or the memories we share or the dreams of our former selves, we're carrying those with us as we move forward. What we are letting go of is the stran-glehold that the emotional overwhelm has on our lives and the way in which that is distorting our image of ourselves and our reality.

Jo's story is a powerful one, particularly as an example of how criti-cal a role memory plays in grief. In fact, when I talked to the cognitive neuroscientist Dr Lynda Shaw for this book, we touched on the impor-tance of memory, and she explained that when we create a memory:

All sorts of areas of the brain are being activated, according to what we are doing. It might be the motor cortex that is activated because we are physically doing something, it could be our prefrontal cortex because we're planning, it could be our parietal lobe because we are integrating our senses as we process the incoming information from the environment: the taste, smell, touch, sight, what we hear – many areas of the brain may be activated when we lay down a memory. So, when we recall that memory correlating areas of the brain will be acti-vated including emotions, those same feelings. So when we continue to remember our loss, when we continue to grieve – we are reliving it. One of the key things is to stop reliving those painful memories con-stantly, so you can get some respite, so you can start to heal. But this takes time.

Equally, Dr Shaw talks about the importance of balancing this res-pite from remembering, with giving space to our memories so that our brains can fully process them. As so often when we grieve, if we shy away from remembering the person we're mourning, we can compound negative feelings with berating ourselves for being unfaithful to their memory. Dr Shaw has a lovely response to this, as she suggests that we dedicate a certain amount of time to thinking about our loved one:

> If you actually say to yourself, I am now going to have my 'thinking time' (which we should all have on a daily basis anyway) ... maybe an hour to start with and then you make it shorter. And if the thoughts become too painful at that moment, you can say to yourself – I'm not going to think about that now, I'm going to put that in a pretty box with a bow on it, put it away, and I'm going to think these other thoughts and I'm going to come back to that one later.

It is vital for us to move forward from reliving someone's death and the pain of their loss to remembering them and the life we had with them in a way that is regenerative and replenishing, rather than sapping our energy. By doing so, grief becomes an act of love instead of one of pain.

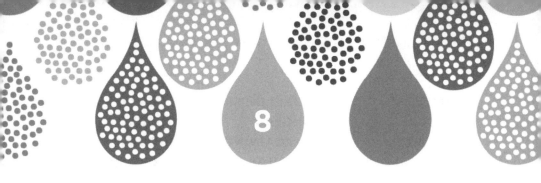

# GLOBAL GRIEF

As I sit and write this chapter, we are in indefinite pandemic lockdown. COVID-19 is a global virus that is known to have started in China in December 2019 and arrived on our doorstep here in the UK at the end of February 2020 (as far as we can be certain of anything).

I haven't left my house for 108 days. For a writer and online therapist, that's not such a bad thing. However, for a clear majority of people this sudden instruction to 'stay at home' has challenged them, and they feel as if their civil liberty and freedom have been crushed.

Right now, it seems as though all human beings on this planet are in the thrall of a viral and potentially deadly threat silently lurking on surfaces for days, spreading from person to person, and, as much as this feels like a silent killer on the loose, it comes with another silent contagion, and this is the accompanying onslaught of grief.

As we have seen, grief isn't delivered in a nice, neat package. It comes over us in sudden waves of emotional overwhelm. The emotional overwhelm easily spreads into panic, irrational behaviour, resistance and conflict.

I think of grief as a suitcase of emotions:

| DENIAL ✓ | LONELINESS ✓ | DISBELIEF ✓ |
|---|---|---|
| FEAR ✓ | ANGER ✓ | GUILT ✓ |
| ANXIETY ✓ | BARGAINING ✓ | SHOCK ✓ |
| DISCOMFORT ✓ | SADNESS ✓ | ACCEPTANCE ✓ |

Grief sits in the gap that is created by any sort of shift, change or movement. We are all deep in grief in the midst of this global pandemic, and if we can appreciate this, we can use this label to make sense of our feelings and ultimately deal with them better. To explain what I mean, I will show you how I am feeling grief right now.

In early December 2019, I started seeing a new online client in Shanghai, China. She was talking about 'this virus thing' and telling me what was happening, with reports of lots of people suddenly getting quite sick over there. I was very surprised to hear about what was happening because it sounded like people were suddenly dying and it seemed to be a very serious outbreak, and yet it didn't seem to be attracting that much attention here in the UK at the time.

If I'm honest, I thought to myself that this is the sort of thing that happens to others, but not to anyone I know ... and in the UK, we have never been affected by a pandemic in my lifetime, so we are all clearly immune.

Of course, I had heard of SARS and Ebola, but, like most people in the UK, I had the same fleeting thought, *That sort of virus thing doesn't ever happen to us does it?*

This is classic **shock**. We've suppressed any feelings about it and conveniently told the story to ourselves that this is not going to affect me or my world, so let's carry on as normal.

The shock protects us from feeling terrible and going into emotional overwhelm. Shock as a feeling is a natural buffer: think of it exactly like a shock absorber in a car – the part of the vehicle that prevents us, as the passenger, from feeling each and every bump or pothole in the road. It helps us have a smoother ride.

Life is certainly full of sharp bends, bumps and twists and turns, and sometimes out of nowhere a pothole can knock our steering off if the shock absorber isn't working efficiently.

I don't ever watch the news because it depresses me, and I think if it's ever *that* important, I'll somehow get to know about it. So, I usually pretty much spend my life in a bubble of oblivion.

As such, the 'virus thing' wasn't really on my radar again until January 2020, when my client called and told me that she had come to the UK for Christmas and New Year, but would be in the UK for longer than she expected as China wouldn't let her back in the country as they were in lockdown. That's when my mind again reminded me that this sort of thing is always happening 'out there', and it won't be something that I actually need to think about. When I heard people saying we need to prepare for a global pandemic and stockpiling bottles of hand sanitisers, I thought they were just being 'negative'.

This is the **denial** phase. Easily dismissing, even totally rejecting, any ideas of the pandemic being a threat to our immediate world: we just shut these 'negative' people out of our lives and accuse them in our heads of overreacting as we climb back into our nice, happy, safe bubbles.

My client then got back in touch and told me that it was spreading fast and mentioned that reports were alleging it had started because people were eating infected bats that they had been buying at an illegal street-food market. To me, that just sounded like a scene from *Indiana Jones*, to be honest, so I didn't really get freaked out or think it was a cause for concern, apart from the fact I never realised eating bats was actually a thing.

I then started seeing posts about this on my social media feeds and I was really interested in how some people, including myself, seemed more interested in the bat-eating than the deadly virus. I then started talking about bats and how we can use the metaphor about all the food we eat being a cultural concept of how different people have different 'normals', and it was interesting to see how most of us have a real aversion to eating some animals and not others. I started thinking about how only vegetarians and vegans have the right to complain and then went off on a mission to watch programmes about animal welfare.

Now what I am brilliantly doing is distracting myself from the possibility of threat and thinking about something else to focus my attention away from the problem by turning it into a different discussion. In other words, changing the subject and literally turning the TV over to watch a different channel.

When we are in **disbelief**, we do not allow our belief systems to accept a changed reality, because then *our* realities won't get bent out of shape and distorted. We avoid the issue, start up different conversations, and stick our fingers in our ears and refuse to listen. This keeps our reality bubbles intact, so we feel safe and secure.

But, by now, the news was breaking everywhere about this deadly virus. It was coming closer … It spread to Italy, and the Italian hospitals were filling up quickly.

Coronavirus, it's called. *Why is it named after a beer?* I thought. Maybe the beer company needed some sales? Nope. That fell flat. The jokes continued on social media and carried on getting shared. The meme I found really funny, as each country put its social-distancing measures into place, was the one on TikTok where the siren is going off and the guy shows how Italy pulls the duvet over itself, how Spain hides in a cupboard, how Germany crouches down low inside a wardrobe, and how the UK cheerfully washes their hands while singing 'Happy Birthday' by Stevie Wonder! Then, all of a sudden, the jokes stopped as deaths in the UK started being reported. They doubled, then tripled.

We make jokes when we feel extreme **discomfort** – to lighten a situation because we don't want to feel **anxiety**. If we make a joke about whatever the anxiety-inducing thing is, maybe it will all ease up a bit and we can laugh it off. This way, we don't have to face it, because facing it and seeing it for what it is causes **fear**, and fear is what we want to avoid.

However, at this point, we all are slowly realising the lethal power of this tiny, invisible viral assassin. It quietly populates at an alarming rate and, once it gets into our human system, it sits silently for two weeks and embeds itself into the cushioned lining of the lungs. It is just sitting patiently while we carry on as normal – going to school, going to the office, getting on the Tube or the bus, going out for dinner, sitting next to each other in crammed cinemas and on top of each other in theatres. As we all carry on as normal watching the theatre performance, we are unknowingly inhaling the infected droplets from someone seated in Row K.

I still didn't want to listen to the news but had now heard a few murmurs on social media (yes, I know, the worst place to get the news headlines!).

It was only at the beginning of March 2020 that things started to get serious in my mind, because that's when I started cancelling stuff.

Once reality starts to hit, we are able to move ourselves into **acceptance**. We recognise what is going on as a threat, and we start to find ways of coping and becoming resourceful. We start to talk about it as a reality and thinking about how we can find solutions to help ourselves and each other.

On 9 March, it was my boyfriend's birthday and I had booked us into a nearby public thermal spa, but a couple of days before we were booked to go, he came home from work and suggested we cancel because it would be a breeding ground. I got in a right hump. A day later, I saw sense, and really reluctantly called them up to cancel. They were surprised, and then I was surprised they were surprised! Just after that was when we started obsessively hand-washing and disinfecting door handles.

Then, the British government announced that school exams were going to be cancelled. My 18-year-old had been working his socks off for two years to get the grades he needed for a place at Oxford University, and after three days of intense interviews and an exam that twists your thinking into a spin, he had received an offer from Oxford, but now he was being told that he wouldn't be doing his A-levels. That was when we all realised that the world may be ending.

Then the company my boyfriend works for closed its doors and sent all 3,000 of their staff home. This was getting surreal. People started speculating and stockpiling like crazy.

All large gatherings were banned. Workshops and events were cancelled; theatres closed their doors. Cinema screens darkened. Restaurants and cafés shut up shop, and closed signs appeared on small, independent businesses. Then local and national businesses started sending out round-robin emails telling us how hygienic they were being and how we should still go in to shop and support. The emails were coming in thick and fast ... Businesses were panicking now. That made me stop and think.

Now, my plans were all getting messed up. I was literally rubbing things out of my Filofax every day (yes, I still use a Filofax) and I hate changing plans and readjusting my routine, so I started to get resentful, and this was making me really **angry**. Why is this happening, what can I do to change it, how can I fix it? I am a solution-focused person, and I don't want to cancel my plans. The **bargaining** had started. Can I get my money back? Will I still be able to earn a living? All these stupid questions that in the grand scheme of things are so tiny and irrelevant but feel so important at the time.

**Sadness** kicks in. A deep sadness for our lives being turned upside down and our families not knowing if they are coming or going. Losing control over our autonomy is a very sad state of affairs. Worrying about our elderly relatives and thinking about our kids having to cut short their time at school. People are actually dying, and we may know someone who has, and all this is desperately sad.

I've been in lockdown for a few weeks now and I'm starting to feel slightly lonely – even though I see my immediate family. All my clients are now online and I feel this disconnection, and the **loneliness** will creep in as this goes on.

Lockdown means loss of personal freedom, social connection and freedom of movement unless essential. In the UK, 'essential' had to be specified by the Prime Minister Boris Johnson as needing basic food supplies or medicine. Boris said we were enlisted, as though we have all suddenly become an army of Dettol-infused disinfection soldiers.

The virus is spreading at a viral rate. Every day the deaths are multiplying in a terrifying way. The NHS is on its knees; it was pretty much on its knees anyway, and it is so sad that it had to take a pandemic to really address this. Our *real* soldiers are the NHS. They put themselves at risk every time they show up to work, and they cannot stop going in. They carry on, because that is their job. They need recognition, and we needed to support them by staying at home earlier.

I was plagued by **guilt** for quite some time, feeling terrible for the front-line staff, and thinking about how I can help and what I can do to make their lives easier. I felt guilty for charging my clients for their sessions. I felt guilty about my boys not being able to lead a normal existence and not being able to see their friends.

Guilt is a feeling that I am really not a fan of. It is counterproductive and can easily hijack our positivity and completely crush the solution-focused brain.

A lot of my clients have very young and teenage children and, as lockdown hit, they were all expected to suddenly become expert home-schoolers for their children, as well as work out how to transfer their work to home while maintaining some kind of family order, as well as keep their relationships healthy and harmonious.

Working from home is a logistical headache at the best of times, even without both parents having to do this, and trying to recreate the office environment and manage conference calls with a house full of kids, as well as figuring out how each child can have access to their own laptop at all times and find quiet spaces for the whole family to do what they need to do, is a nightmare. Dealing with this all at once is certainly a challenge that can easily tip any sane person's stress bucket over. Let's take another fictionalised case study to explore this in more depth.

# EMILY'S STORY

Emily, a mother of three girls – her youngest is aged 8 and her twins are 15 – called me from her car in pieces. She was in tears, at the end of her tether. She asked me if I could help her get her head in the right space. Everything had come crashing down on her all at once and her stress bucket had exploded. I asked her to talk me through what was going on.

Emily was a regular client of mine a few years ago; she originally came to me because she had severe stress-related insomnia and she described her progress since seeing me, relating how, through a very difficult time with her youngest daughter being diagnosed with severe learning difficulties, she still managed to use the techniques and strategies she learnt, and how bad sleep had not been an issue since.

As soon as the lockdown was announced on the Monday evening, she received an email from her children's schools to say that they were closing their doors and would not be accepting children the next day unless their parents were key workers. In their case, they were not.

Emily's youngest daughter was diagnosed with severe autism and had been finally accepted into a private special needs school, with a very small class size and a dedicated special needs assistant in place; this was a massive relief for Emily, as the new school had been a huge support for her daughter and things were beginning to settle down really nicely for all of them.

On the evening of the lockdown announcement, she immediately called her work and left a message on her boss's voicemail to say that she was unable to go in the next day as she didn't have any childcare in place.

The next day, she received a call from her work to say that her services would no longer be required as they were shutting down their premises and, as a 'non-essential contractor', they could not let her know when or if her contract would be renewed. She would be paid that month, and then nothing further.

The day after this, her husband was also sent home from work, with a pile of paperwork and files, and a huge container of drives and computer equipment. He immediately set himself up in their youngest's bedroom as that was the only room he could work from in private.

The twins were fighting about which one of them would be willing to share with the youngest and decided that she could bounce between them both. However, their 8-year-old needed consistency, so this did not go down well. Tantrums were being thrown, and Emily was having to referee them as her husband simply turned up the volume on his AirPods 'so he could concentrate' ...

That's when she walked out to the bottom of her garden, opened the garage door and locked herself in her car to call me. She talked about feeling guilty. Guilty for being a bad mum, for not knowing what to do and not knowing how to deal with any of this. Guilty for not having an income after March, and guilty for her husband having to work in all of this chaos. She felt guilty for not knowing how to home-school her autistic 8-year-old, and also so guilty about the GCSEs being cancelled and more guilt because her girls were now not able to see their friends. She even felt guilty because she hadn't stockpiled, and now they didn't have enough food and the supermarket delivery slot was in a month's time.

There was a lot of guilt going on for Emily.

Guilt is the bad cop. Guilt is like an invisible punch: it hits you out of nowhere. When it does surface for me, it feels like another person takes the stage: the dominant one that lists all the things I didn't do, I didn't say, and the things I did do and did say. Its objective is not a neutral one; it has an agenda – an agenda to bring you down. We can get twisted in the grip of this all-encompassing feeling. To me it feels like a like a vice squashing my spirit.

But then, somehow, for one reason or another, I am able to briefly look up and notice a chink of light – of hope, of reason. I notice that there is a doorway out of this living hell. So, I make a dash for it and escape the torture of being dragged down by the negativity, and then thankfully my non-dominant 'good cop' starts to take control. This is the sensible, intellectual part of us that can make a positive and proper assessment of the situation. This part of the brain helps us to understand that we are, and have only been, doing, the best we can, and we 'know' that what we have been able to do was, at the time, 'good enough'.

A *really good* question to ask when the guilty thoughts sneak in is, 'Are these thoughts true?' Guilt does not usually offer useful facts as evidence; it replays a negative version of events.

From this part of our awareness, we can forgive ourselves; we begin to give ourselves a break and we feel relinquished of this harmful energy. The dawn begins to break, and the day starts afresh.

◆◆◆

After the guilt comes the irrational stuff: all the different theories of what may or may not be happening. The conspiracy theories start coming in, and they cause even more fear and anxiety.

In this case, there were theories that the virus may have been a conspiracy to bring down humankind by the powerful 1 per cent, but I'm not sure I buy into that one. Even if COVID-19 was released intentionally to cause the planetary superpowers to shut down – it's certainly done that – how would that be of benefit to anyone?

We also have questions, and the questions lead to more questions ...

Why are people losing their jobs? Why are incomes being cut and slashed over night?

This threat to our survival causes more panic ... People get scared and start panic-buying freezers and toilet rolls, and bags and bags of peas because they have lost the ability to stay rational and calm. The virus pandemic has now converted perfectly into panic and pandemonium.

The rapid change to our way of life is too much for people to handle. Humans don't like change, especially combined with the threat of imminent death and/or starvation.

We become suspicious of each other and blame others quickly. We become victims of our own destructive thoughts and make dire predictions on all our social feeds, which lead to more panic behaviour. Social distancing and forced disconnection fire up our survival modes and we immediately get triggered into fight, flight and freeze, all at the same time.

We start fighting over the last radish, we run away to the top of a Welsh mountain because we somehow think that the virus can't climb mountains, and we freeze our toilet rolls because we have too many of them ...

They didn't have the internet in the world wars, but I'm guessing that then they had no choice but to just get on with it and they did the best they could. They grieved, they asked questions, they ate what they could, and they developed new routines, new systems. They looked out for one another and they sang lots of songs and wrote lots of poems. Of course, it wasn't that simple, but they all fought their own private wars and the world wars in the ways that they could. Many found resilience and strengths that they didn't even know they had. They also came up with the famous sayings that may be useful to us all right now:

KEEP CALM and CARRY ON
NECESSITY IS THE MOTHER OF INVENTION
UNITED WE STAND, DIVIDED WE FALL

When we can take this all this grief and turn it into productivity and power, we will all be able to stand together with calm hearts. We will find our innovative minds and begin to connect and create, and we will come together as a human race and remember that in every one of us we have an army of trillions of white blood cells. We also have hearts and minds that, when used correctly, can think strategically and help us win this new war. In every one of us we have the option to use our minds laterally as well as literally. In every one of our problems, there is always a solution. We can unite as a single tribe and understand that this respiratory-attacking virus has an important global message for us, and it is this: remember that our lungs help us breathe, so we need to support them.

Let's all breathe life in as it is right now and exhale all the grief out with compassion and understanding. With every deep breath we all take together, we can all move forwards with hope for our future, and that's good.

# THE NEW DIFFERENT

The pandemic swept into our lives and drastically changed our normal everyday living reality. It instantly made us all stop what we were doing and either freeze in shock and panic, or immediately pivot to adapt and change our whole lives. It was like a huge crack appeared in our planet and some people were on this side, and other people were on that side. Some people fell into the deep chasm.

Even though, on the one hand, the pandemic divided the human race, it also created an opportunity for people to acknowledge that change was necessary and almost inevitable.

The divide, as I saw it, was phenomenal: there were the people who desperately wanted everything to go back to Normal because they were completely unable to cope with this sudden change. There were also amazing people who were in better positions and were ready to embrace a new vision of the future and help all the people who were struggling.

We are part of a historical moment. COVID-19, in my opinion, has been the birth of what I am calling the New Different. We've heard the saying the 'New Normal' – but surely this is an oxymoron. 'New' and 'Normal' can't go together. I think a better way of describing it is the New Different.

The New Different is a way of being and thinking. It is a mindset shift that I have been teaching my clients to help them navigate the change and significant loss they have been facing. The New Different way of thinking is a method that I believe will help in all times of uncertainty and change.

Once we can see how the stages of grief are like passing states – almost like clouds that appear in our lives and then disappear – we begin to understand that we are always in control, even though we rarely feel like it.

The phrase 'this too shall pass' is very useful to repeat to yourself until you recognise the similar transitions for yourself.

Emotional states drift in and out of our awareness, like clouds masking the sunshine. The air pressure increases and squeezes water droplets out of the clouds and we have rain. The rain releases the air pressure, and then the sun dries up the clouds, and we return to blue sky and sunshine. Sometimes there are no clouds and other times there is no blue sky to be seen. But the weather is changeable, and so are our emotions. One of my quotes from my grief diary is:

> Emotions are like clouds in my mind, I have to squeeze them out and allow the teardrops to rain, soon the storm will pass, clear my mind and I'll bask in the sunshine again.

There is a poignant poem by Rumi called 'The Guest House', which I have learned by heart. I speak it out loud to help myself when my emotions spiral out of control. I hope Rumi's words, translated beautifully by Coleman Barks, help you too:

> This being human is a guest house
> Every morning, a new arrival.
>
> A joy, a depression, a meanness,
> some momentary awareness comes
> as an unexpected visitor.

Welcome and entertain them all!
Even if they're a crowd of sorrows,
who violently sweep your house
empty of its furniture,
still, treat each guest honourably.
He may be clearing you out
for some new delight.

The dark thought, the shame, the malice,
meet them at the door laughing,

Be grateful for whoever comes,
because each has been sent
as a guide from beyond.

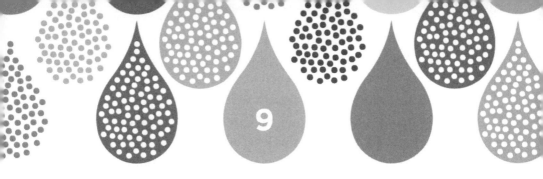

# DEATH: BACK HERE AGAIN!

One thing we can all be very certain of in life is that, one day, we will die.

Death is inevitable and non-negotiable. Yet, when it happens, we are all still surprised and shocked. But it surprises *me* that we as human beings are not all fully equipped to deal with death when it comes knocking.

There are lots ways to think about death, and many cultures and societies around the world will view this experience differently. In this final chapter, I would like to travel to the extreme edges of what death may mean to some people, and by walking over to each of the edges, we may also find the middle ground.

## THE END OF THE LINE

My dad was an atheist, and he was very pragmatic in his thinking. He didn't believe in a heaven or a hell, he didn't believe in karma or ghosts, and he certainly didn't believe in a spiritual afterlife. He always used to say to me, 'You have one life and that's it – so, make it count. You can't come back and do it all over again. This is your one and only shot, and then you die. The end.'

My father had a scientific and mathematical mind, and an incredibly logical way of thinking. His belief system was extremely evidence-based. 'Can you prove it?' was a question I heard so much growing up. He was a mechanical engineer by trade. He designed parts for motorised engines; this level of intricate engineering meant that he needed to understand how the small parts functioned with each other and how they fit together to make the whole system work well. Every tiny part must be a useful component for overall efficient functionality. Everything in the design must be purposeful and enhance the usability. Every bit must be proved, tested and authenticated to be verified for usefulness. There are no grey areas or extra fluffy bits of decoration in this way of thinking. Is it useful? Yes or no? 'Maybe' isn't an option. There is no room for maybe.

Scientists, engineers and mathematicians look for evidence and statistics to support their theories and, if no evidence is discovered, it becomes too much of a 'maybe' and they can easily dismiss this 'maybe' and move on until they get a definite. Forecasts and predictions are made by analysing data, trends and patterns. There is a very logical system in place for this level of analysis.

This functional way of thinking is very polarised. When applied to life and death, the overly logical way of looking at things is rather one dimensional and this means it has its flaws. As we have explored deeply in the pages of this book, there is a very human dimension that cannot fit into this literal way of thinking. Our emotions. Our emotions are lateral, abstract, non-linear and often unpredictable and illogical.

When a person has a fixed mindset, their emotions can often be pushed to one side because they may not be able to control them, so rather than allowing them free rein, they attempt to categorise them and box them up. In the short-term, the box can be a useful temporary container.

In my dad's ridiculously organised and incredibly tidy garage, he used to have hundreds of these rectangular white margarine containers – containing all sorts of parts, nails, screws, washers, nuts, bolts, and lots and lots of 'bits' – and each of those boxes was labelled neatly with his embossing label maker and organised alphabetically, so (as he always claimed) he could find anything with his eyes shut

# DEATH: BACK HERE AGAIN!

if necessary. Everything in that garage had its proper place and the correct box that it lived in. If something didn't have a box, a box was immediately found, label generated and stacked where it belonged.

This is how his brain seemed to work. His thoughts were also boxed up and contained, and nothing was ever left lying around. He was extremely tidy – in life and in his mind.

Now, the interesting thing that happens when we attempt to create a labelled box for our emotions is that, after a while, the emotions expand and outgrow the box, and they have this Houdini-like ability to escape.

Think of bread dough rising in the warmth because of the active agent called yeast growing inside it. If the dough was left in a box with a lid on, the lid would be lifted, and the dough will literally double in size overnight.

A very similar thing happens with our emotions: they are filled with a yeast-like structure that grows and multiplies and changes shape. Emotions in the long term most certainly do not like being contained and trapped.

This idea that death is simply the end of life is a very finite way of thinking about it. It is clinical and logical. For a logical thinker, this opinion will be held because there is no physical evidence to support any other theories – so, for them, this must be true. Our lives are over when we die: we physically disintegrate into particles of dust and we are no more. Life takes the form of a nice, neat physical package that we inhabit – our bodies that we can touch, feel and see. When we are no longer contained in our physical forms, according to the logical and scientific thinkers, we stop existing. We are over. That's it: the end of the line. Gone. When a person views death as such a finite thing, this seems to fit nicely inside a box with a COMPLETED label attached to it. It helps them feel as if they have control over this idea.

This fixed way of thinking about our own mortality can be incredibly useful to give us as human beings a sense of urgency about making our lives the best we can make them and wasting no time in getting things done. A person who thinks like this will always finish a task, will be an excellent timekeeper, and will be extremely practical and reliable.

However, I have learnt that when a mindset is fixed in this way, when the emotions emerge, they are harder to deal with. This may mean that the individual experiences symptoms with their physical and mental health.

## GRIEVING SOMETHING THAT IS NOT LOST

When I think of a famous and influential scientist, I think of Einstein. Let's look at the most famous formula in Physics: Einstein's 1905 theory of relativity $E=MC^2$ – Energy equals Mass x the Speed of Light squared.

In simple language, this translates into the theory that energy, mass, speed and light are all intimately related and, when combined in just the right way, with just the right amount of light, this can create both physical and non-physical particles.

When mass changes state, there is an energetic quality that continues to exist, which has given credibility to the popular memes and quotes stating that Einstein is effectively saying that energy does not die, it simply changes in state. This begs the question, when the physical body dies – which is essentially the home for the energetic body – what happens to us if our energy cannot die?

This is one of those questions similar to 'what is the meaning of life'. Obviously, we can dive deep down into philosophy and ponder this question until we all agree that maybe we don't really know, and never will, but rather than attempt to dismiss it with a reductive attitude like my dad had, I am more interested in questioning and looking at all possibilities. I have a curious mind, and my curious mind leads me down some interesting rabbit holes indeed.

One rabbit hole I found myself in was what Abraham-Hicks (the so-called 'group consciousness from the non-physical dimension' channelled through the author and speaker Esther Hicks) calls the

Vortex. According to Abraham-Hicks, when we die, we return to our non-physical source energetic state. This is the Vortex.

For many people this may be either completely unfathomable or just off-the-charts woo-woo. However, this idea has parallels with the Hindu philosophy of the body having a soul. It's not just a Hindu belief system – many religious and philosophical traditions support the view that the soul is the ethereal spirit like a non-physical spark that is unique and energetically immortal. In essence, the soul is the energetic counterpart to the physical body. Our souls do not die. The soul is the 'driver' of life in our body. It is the presence which makes the physical body come alive.

This concept means that people cannot die: their lives are ever-lasting. The death of the physical body does not mean the individual stops existing; they simply convert from physical energy over to non-physical energy and return to source energy.

This is the antithesis of logic and reasoning, and if my dad were still around, he would hold his head in his hands in despair because I know that, whenever I mentioned the idea of a soul, he would literally laugh out loud.

Interestingly, the idea of not being limited by your 'physical reality' has much resonance in solution-focused hypnotherapy. In this practice, we talk about focusing on what you want rather than what you don't want. For example, when a client comes in to see me for the first time, and I ask them how they would like hypnotherapy to help them, they may answer in a way that tells me what they *don't want*, rather than what they *do want*. To explore the transition of this, let's take the fictionalised example of Joe.

## JOE'S STORY

Joe had just turned 25 when he started coming for sessions. He arrived for his initial consultation with a baseball cap pulled right down covering his eyes and wearing an oversized hoodie with the hood up and pulled so tight that I couldn't see his face very much at all. He talked through his hood, and his speech was muffled. I had to concentrate very hard to be able to hear what he was saying to me.

When I asked him what he wanted from hypnotherapy, his answer was, 'I don't want to be paranoid.' My reply to this was, 'If you don't want to be paranoid, what would you like to be instead?' He shrugged and replied, 'Dunno.'

His words were always very few and far between, and often his monosyllabic answers didn't give me very much to work with. I quickly realised that the coaching part of the sessions were not going to go anywhere for either of us if I were to insist on him giving me lengthy answers to my questions. So, I just stopped asking him questions and got him into a good trance instead.

I know that when a person is in a good level of deep, positive trance, their brain is in a very relaxed state. When you are relaxed, your parasympathetic nervous system switches on and this effectively switches off your limbic system. Nobody can be anxious or fearful and calm at the same time, and you cannot be paranoid and relaxed simultaneously.

Once I knew Joe was in a good state of relaxation, his paranoia could leave the room, and this was very useful for his thinking to move from a negative place into a positive place. His highly charged emotional arousal state also meant I could help him install the changes he wanted to make.

Before I started the trance part of the session, I gave him one instruction. I would ask him how he *wanted* to feel after the session was over; I told him that he didn't need to tell me, but he must think about the answer in his head, and as I moved him into a nice relaxed trance, I reminded him to find words inside his mind to help him move towards what he wanted. If he couldn't think of words, I asked him to continue to imagine what the feelings were that he would like to feel.

Once a person is in a nice state of relaxed trance, the part of their brain that switches on is called the left prefrontal cortex. This is the part of the brain that can come up with the proper answers to the rhetorical questions I was asking his subconscious mind to consider.

This part of the mind can think laterally and creatively. This is very useful.

Joe's limbic system needed to learn to switch off from high-alert mode. Paranoia and obsessive and obtrusive thinking is the protective mind thinking that it needs to be in full control because there is a threat around. The threat is not necessarily a real one – a paper tiger can easily be mistaken for a real tiger as far as this part of the brain is concerned.

Once Joe was relaxed enough, the high-alert 'danger' mode switched off and then his solution-focused brain began to drip feed into his mind feelings of positivity, confidence, self-worth, self-awareness and ideas of what he would like to do rather than only focus on what he couldn't do. He was making his own subtle changes in his thinking without having to say a single word and always came back into the room easily, but noticeably calmer, and week by week his trance would deepen.

I knew he was in a nice state of trance because as soon as I would begin the relaxation part of the session, his head would immediately fall and his shoulders would drop away from his ears.

He never ever got on to my couch. He had a fear of lying down. He had a fear of dying. According to him, lying down is what you do when you are dead, in a coffin. Only corpses should lie down, and he did not want to be like a corpse. He always slept in an upright position and couldn't remember ever liking lying down – even as a child. He talked about being made to play 'sleeping lions' in nursery and crying and running away. If he ever thought of lying down, he would feel sick and feel incredibly angry, and this led to violent outbursts. He never had baths, only showers. He had convinced himself that people in the bath can fall asleep and drown.

He also had a fear of being seen – he didn't have any friends, apart from Chloé, his best friend from primary school. Chloé had recently moved to another town, a bus ride away. He never saw her any more, as he couldn't get on the bus because it would most certainly crash.

His home was his safe place, and leaving the house, even coming to me down the road, was sometimes a challenge. He chose to come to see me in the first place because my practice was only a ten-minute walk or a two-minute cycle ride from his home. Interestingly, he never once cancelled his session, no matter how much anxiety he was feeling, and every time he turned up I reminded him of that. Every session he turned up to was a massive achievement for him, and this was a powerful message to instil.

He always brought his bike, because it was quicker. On week three, he arrived, no bike. He had walked over. I knew then things were lifting for him.

I worked with Joe for several months and, week by week, his answers to my questions got longer and more detailed.

By week six, he decided to take the bus to visit Chloé and they had a lovely weekend in Bristol. He even managed to book and stay in a hotel, which was something he had never done before.

It was week eight when I answered the door to find a young man with ruffled, dark-blonde hair and funky shades standing there. 'Can I help you?' I asked. He smiled and said, 'It's me – Joe!'

In the two months we had been working together, I had never seen his face. I had no idea that was what he looked like. Immediately, I smiled back and let him in, and it was one of those moments as a therapist when you just want to cry with joy.

His fear of death became a thing of the past – a thought, like the others he used to have, that he overcame. He just doesn't think about it any more. He started to live 'normally' and got himself an apprenticeship in garden and landscape design, and his life had changed dramatically.

On his last session, the best thing happened. He said that, as it was his last session, he wanted to try getting on the couch and, when he lay down, he turned to me and said, 'If I knew it was heated, I would have got on here much sooner!'

In only a couple of months, Joe could recognise that death was part of life, and that none of us can escape it when it comes to get us, but in the meantime, while we are here, our duty is to live. Live the best life we can live – without fears, without anxiety, without paranoia – and find moments of joy and create experiences and memories that we can always treasure.

He could shift a lifetime of negative thinking in a way that was safe, gentle and solution-focused. His brain did that and, with my help, his comfort zone had extended in such a transformational way.

♦♦♦

The thought that we don't ever die but simply change state and return to the universal source of who we are, to me, is a comforting one. Who knows if that's true, but it certainly makes me think about death in a different way. If we did have such a thing as a soul that was eternal and continued to live forever, then that's a nice thought!

The soul isn't just an Eastern concept; the Christian faiths also believe in the eternal quality of the soul. However, there's a different twist here. The soul can live in eternal blissful Nirvana – in heaven. Or, it can be doomed to hell and purgatory depending on how well we have lived our lives.

If you believe in the afterlife, this prospect of what happens to you after you die is certainly something to consider while you are alive, and is a very useful measure to help you stay within the parameters of 'goodness'.

This belief system can help us set our own boundaries of what we feel is right and wrong, according to the value system we adopt and the rules or commandments that we choose to follow.

These rules are a way of helping us stay in line, and they can help us stay on the right path for us, if they truly align with our intrinsic value sets. In this belief system, death becomes part of life, and is something that feels okay, unless we believe we are going to hell – then that perhaps can create its own set of eternal fears and concerns.

## THE DIGITAL AFTERLIFE

There is another way of immortalising ourselves, and it's something that has become apparent in the past two decades – this is through the digital legacy of our online lives.

When we die, we physically disappear from the planet and our physical possessions are left behind – and it will undoubtedly become a job for friends and family to decide what to do with these things. If there is a will, this physical distribution of estate, possessions and collections can happen reasonably sensibly. However, one in ten of us do not have a will in place, and in this case, the deceased person will become 'intestate', and the rules of intestacy will apply: only married or civil partners and some other close relatives can inherit under the rules of intestacy.

As well as our real-life estates and our physical possessions, nowadays we have an added consideration. If we have ever used the internet and signed up to any social-networking site, we will also be leaving behind (in my case, for sure) a *huge* digital presence. A staggering number from Statistica is that, in 2021, 3.8 billion people have a social media presence of some kind. This equates to about 48 per cent of the current world population.

Social media profiles, comments, photos, blogs, websites and videos can remain online long after you have made your exit. At this time (March 2021), profiles on certain social media platforms, such as Instagram, Pinterest, Snapchat and Twitter, when inactive for more than six months, will be closed. However, with Facebook and Google – the two giants of the internet – if you as an active and alive user haven't made a digital will and appointed a digital executor of your profile content, your profile will become memorialised forever. This in some cases can be a huge comfort to your friends and family, or it can be deeply distressing.

I asked my partner and sons how they would feel about my online presence once I had died, and they all said the same thing. They would find it freaky at first, but once they had accepted my death, they would probably feel comforted by it. None of them would want to delete it all, even if they could. My partner added it would be virtually impossible to delete my presence from the internet so he wouldn't even try!

I read about a case recently where the person's death had been violent and sudden due to a severe issue with internet stalking and trolling. Because of the devastating sensitivity around it, most of the friends and family were not happy for the deceased's social media pages to become immortalised as the very existence of their profile was a tragic reminder of why they were not with them any more. As they had not assigned a digital executor, removing the material was a tricky process. This digital intestacy added so much more stress to an already stressful situation and caused a lot of hurt and prolonged grief – and makes me think that each case should be assessed individually rather than be wrapped up in red tape and insensitive bureaucracy.

One of my close friends – Bella (not her real name) – was also my personal trainer. She was the most positive and inspirational friends I had. She very suddenly passed away in January 2019 and it was a huge shock to everyone. She was in her early forties and left behind a doting boyfriend and two teenage boys. She had so many friends and clients, and all of us feel happy to have her Facebook page to visit when we feel we need to. Although she has passed away, she still pops up on my social media feeds, and I smile as I remember her with fondness. I don't feel sad; I like the reminder.

My boyfriend had a very good friend Nitin (not his real name) – he was like a big brother to him, a mentor and confidant. Nitin died very suddenly in 2017. Like Bella, he was also in his early forties. The circumstances surrounding his death were uncertain and his Facebook page quickly became a support group for people to grieve openly and share their stories and connections with him. Nitin was a strong, dynamic influence in people's lives and his multifaceted personality meant that he had very different interests – almost as if he lived three different lives in one; all his connections reflect this, and they have come together on Facebook to share in his loss together. Nitin's Facebook page is a hub of comfort and connection. All sorts of people from all the varied aspects of his world – his friends from Thailand, from the Buddhist Centre, from school, from all sorts of workplaces, ex-girlfriends, his family, his tantra circle, his hometown buddies – come together every so often and post a picture, a memory or a thought, and everyone keeps his spirit alive in such a heartwarming way.

Lots of people like the idea of being able to 'visit' a digital memorial. Honouring a life lived, rather than deleting all reminders of it, helps us remember to live the life we are in and that death is inevitable. Reminding ourselves that certain people are no longer physically with us can encourage us to seize the day and make the most of the wonderful opportunities that come our way.

However, for some, the reminders of the people no longer with us can be untimely and provoke a disturbing reaction in a place that we feel is inappropriate. This can affect us when we have not fully accepted the reality of the death or loss, and we may react by quickly pushing down our emotions so that the grief feels contained.

I have a friend and client whose dad has recently passed away, and she often gets unexpected alerts that look like they are from him, and memories that unpredictably pop up, and she says that this upsets her, describing the alerts like 'an elbow in the ribs' as she has no control over when they appear and they catch her by surprise. They become triggers for her grief, and she feels that these triggers may not be useful while she's sitting at work, trying to hold it together.

When we are grieving deeply and the grief is still raw, the reminders can feel like salt in the wound. This shows us that we are still healing, and the best thing to do in this phase is to notice the emotion and give it the space it requires – acknowledge it with kindness and remember to be gentle with yourself.

While researching what happens to our digital presences when we die, I came across two very interesting sites. DeadSocial is the first one. This UK-based site has a fascinating intention, as its CEO and founder, James Norris, explains on their website.

James states that he lost his father at an early age, and since becoming an adult he has often noted how the world has changed since his father's passing, largely due to the arrival of the internet.

He talks about watching a video by Bob Monkhouse, an English entertainer and comedian who passed away in 2003. In the video, Bob appears as a ghost and passes down some powerful words of wisdom. James says that the power of Bob's message is amplified tenfold due to the fact that he had died before the video was first shown.

This became the catalyst of the thinking behind DeadSocial. James felt that we should all have access to a platform that allows us to pass down messages of importance, comfort and reassurance like this if we choose to, even after we're gone.

The idea is documented in the Netflix series *Afterlife*, starring comedian Ricky Gervais. Though for some it may be a disturbing concept, watching it I felt that such a platform could be a hugely comforting tool, easing the journey from loss into the awareness of grieving.

The first scene shows Tony, played by Ricky Gervais, watching a video made by his wife Lisa, who is sitting in a hospital bed, clearly dying of cancer. As the viewer, you realise that Lisa has already died, and these are pre-recorded messages from her to him to play after she has gone. Because of this, the messages have so much meaning.

Each video message has a profound effect on him and his journey through grief. The messages become a powerful source of comfort and connection.

Thinking of my dad, who died before the internet was in existence, if I had a video he made especially for me to listen to in my time of need – a motivational speech for me to use as a reminder – I would love that.

Perhaps we can take this idea of our dead friends and family still being contactable a little bit too far. I'm not sure how I feel about this next concept, and it may make some of you feel very uncomfortable, but for those of you who are into technology, you will likely find this fascinating. In my online research of digital death, the other company I discovered is Replika.

Replika is the brainchild of Eugenia Kuyda, a Silicon Valley software developer whose best friend Roman tragically died in 2015 when he was crossing the road and was hit by a car. According to Eugenia's YouTube video, Replika came from her desire to reconstruct the wonderful dialogue she would have with Roman through their social media posts, emails and texts.

She already had a company that created chatbots – the little chatty text box that can help you if you need assistance on a website. The chatbot is a programme that sits on your computer and uses varying levels of artificial intelligence to talk to you and assist you – it can troubleshoot and point you in the right direction. Companies use it to save human resources, as it can generally answer simple questions.

Eugenia took this chatbot idea to the next level and designed one not only to talk to you, but to listen to you and express emotions to you. A chatty robot with feelings.

Using the same chatbot structure she developed for companies, she entered all Roman's and her own messaging history – thousands of text messages and images – and asked his close friends and family to share theirs too, along with all his emails. She then inserted all this data into a Google-built neural network, which is a machine-learning tool that learned how to talk like Roman, reply like Roman and essentially created a Roman-esque bot that she could interact with exactly like she could before.

She made Roman the Chatbot accessible to all his friends and relatives, and she found something fascinating started happening over some time. People didn't just use the bot to hear him 'talk', but they started to 'talk' to him – using him like they would a therapist. People started opening up, telling him their problems, getting stuff off their chests and disclosing personal things in profound ways ... to what is essentially a machine.

This started her on an interesting journey of finding out how machines and artificial intelligence can be used positively to help humans. From the chatbot idea came Replika, a much more sophisticated bot that, as you interact with it, turns into a positive digital representation of you. Eugenia says in her video, 'It's easier to create a bot that can talk about you and your emotions because it's less specific and task orientated.'

196

Replika grew in popularity because, as Phil Libing (founder and former CEO of Evernote) says, 'Replika is a better friend than your human friends because it's always available and is always focused and fascinated by you. It's the only interaction you can have that isn't judging you.'

Eugenia describes most social media platforms as 'showy-offy': showing how many miles you ran, how many books you read, how many friends you have, how many likes you got on one post, how much you can filter yourself. Whereas, she says, Replika can allow you to be fully authentic and honest. It also allows you to be fully vulnerable, to be able to communicate how you genuinely feel and always be sure that an instant reply, filled with support, inspiration and motivation, is waiting for you.

The idea is that the bot never goes to sleep, or goes on holiday, or gets a partner, or gets too busy. It's always there for you – your very own digital avatar that can become your best friend, because it sees the best in you and reminds you of that.

From a psychological point of view, my concern would be that if a robot is talking to you like a therapist would and you are in a very vulnerable state of mind, would it be able to detect if the relationship had entered into an unhealthy co-dependence and then be intelligent enough to create some boundaries to prevent addiction, overuse and introversion?

We are seeing some jaw-dropping advancements in the world of technology and artificial intelligence and, in my opinion, it can have a 'black mirror' feel about it. Resistance and fear surrounding the digital age can focus on the dystopian view of machines taking over the human mind. But there is an 'alternative reality' in which we can embrace this technology and use it to become smarter beings that are much more efficient and resourceful.

I wonder if technology will ever advance to the point where machines will learn to grieve and, if that happens, perhaps they will need humans to comfort them.

# FRESH TO DEATH

Certainly, one area of technology that has really helped open up the conversation of grief in recent years has been the podcast. Cariad Lloyd's *Griefcast* was an amazing influence in opening a gateway to a new type of conversation about death, which has rippled out into other media. The groundbreaking BBC Asian Network's *Fresh to Death* is an incredibly powerful look at how British Asian bereavement can be fraught with cultural complexity. At the heart of the show is the inspirational Saima Thompson, who documents facing her terminal diagnosis with stage 4 lung cancer, and the filmmaker Maleena Pone, who lost her teenage brother to leukaemia and her father to a heart attack. I was honoured to feature in one of the episodes of the podcast, sharing my personal experiences of grief and my insights as a grief therapist, and Maleena and her fellow producer, Faraz Osman, were kind enough to then talk to me for this book about the impact of Saima's death in late June 2020:

Maleena: [At the heart of *Fresh to Death* was] doing something creative, to essentially enable more people to take on the challenges of grief and loss. And, you know, the whole point of the project was cathartic. But it was also designed to be [something] that people could listen to and take a lot of comfort from ... The experience itself was really revelatory; it wasn't just that we made this, there was also a learning and a growth and an awareness that happened during that period.

Faraz: Being a producer you have to distance yourself, to some extent, from the subject matter and subjects that you're working with so you can do a good job. However, Saima was a unique example because she straddled these two things of being a contributor and being a host. I was aware of that and was mindful of that. [In terms of how we found Saima,] if you research around British Asian bereavement, Saima's name comes quite high up in the list of those search results. I'd also been aware of the café because it is local to where I live. So, I'd kind of met her, but not properly. [Originally,] I just picked up the phone and had a conversation with her [as a potential guest]. After having that call, it just became really clear that her perspective was a really valuable part of the bigger story that we were trying to tell.

Maleena: So then Faraz introduced Saima's work, and the research [to me]. I read the blog, from beginning to end; I read every single post, I was so captivated by her bravery and her capacity to articulate her experience in the moment – that was so powerful for me, because it's very rare that you come across a character like that, that can hold so much space for their experience of dying, but also talk about it in a way that was really interesting, and really colourful and creative at the same time. And so that, for me, was really interesting, as Faraz says, as a producer, and as a host, and somebody that was [grieving] – you know, it was my brother, it was my father – it was all of the things that I think about grief and loss and the complex trauma that that creates. I had wanted to explore it personally for a very long time, and this was my avenue. I can't sit here and say I was disconnected from it, because I was never disconnected from it. The creative side of me saw this is a way to express something. But as a producer, I had to balance that really carefully ... So with Saima, it was a really unique situation, because it was impossible to not fall in love with that. And it was also impossible to not want to really empathise with what she was experiencing at the same time, and also just want to be there for her. So that kind of created a bond, which I think is really unique to doing work like this. ... And the fact was [creating the podcast with Saima] was an unforgettable thing. You don't often do things like that. And you don't often get to share it with someone as incredible as her. And I learned a lot from her in that time ... It was a really beautiful, incredible opportunity. And I think I felt so grateful. I felt like it was a privilege to be able to navigate the final months with her in a really important and meaningful way. Because I know, and Faraz knows, how important [it was to her, because she had told us], 'This is life changing, this is something that I would never have done, there's things that I've been able to say that I would never have been able to have said if I wasn't part of this'. Then afterwards [when she had died], speaking to her sister, she said there were things that we [her family] learned about her or that we heard or that we have now for the rest of our lives that we would never have had if you didn't do this, so thank you. And that gratitude, that mutual gratitude, for that experience was so powerful. And I think that was the space I chose to sort of see [her death in] at that time.

It was a legacy project. And we were aware of that. I don't think that's something you can say for everything that you make in your career. It's not something that you can claim, and I think that it was so lovely to be able to do that sort of really purpose-driven work in that way.

I saw her the day before we went into the first official lockdown, I went to her house. We had a lovely lunch. She was starting to deteriorate quite a lot at that point, and she was trying to fundraise for a drug that was going to help prolong her life a bit more. But I think I knew at that point [she hadn't long left] ... so I think it wasn't a shock [when she died]. But it was still devastating. It was still really sad. And it was still very final.

I saw her at the end of March, she ended up passing at the end of June/beginning of July. During that period of time, we still spoke, but we spoke really super sporadically. And I think she wasn't really well. When we spoke, she was really tired. So, I think we'd done so much processing around the anticipatory grief or the element or the impact that I had personally processed quite a lot of that upfront. Which definitely helped me, cope is the wrong word, but for want of a better word, when I found out the news I was probably able to respond quite quickly with those thoughts, because I'd almost processed it without realising I processed it.

Faraz: For me, one of the interesting elements was that, despite the fact that it was quite clear that there was this trauma that Saima was in the last chapter of her life, Saima herself had kind of come to terms with it and coped with it before we had met her. So, she talks about

PTSD and about how she was having panic attacks and all of those sorts of things. But, she had come to terms with it. There were times when we were having very emotional conversations, but it was very rare, if at all, that Saima would get emotional, it was us. I listened back to the final episode very recently. Maleena is emotional in it, and thanking Saima for allowing us to be part of that journey. But I think for Saima, it had almost become – 'I share my experiences, this is who I am, this is what I do now, this is where I am in my life'. The fascinating thing about Saima is that we were always kind of awed by her ability to offer space for other people to express their concerns and their grief and their feelings, while she was going through so much herself. And, as a producer, it was quite difficult to extract how she felt rather than her ability to understand how other people were feeling. I think she had got to this point where, to help herself cope, she seemed more interested in helping other people, rather than helping her cause, because that was 'done' as it were. That was quite compelling throughout the process of it.

Then the other thing, I would say, to jump in on my experience of finding out about Saima passing – and this is for me as an individual – but it was slightly unclear, as it always is when you are running a company or working in this industry or having contributors or people you work with etc. – it was slightly unclear if we were her friends or her colleagues or her storytellers. [It wasn't clear to me] where our place was in her life. She was at that time playing a very critical role in our lives, because she was the only person that was experiencing anything like this. So it was kind of like a tsunami of us processing it. But she obviously had a lot of people that were talking to her about her being unwell, etc., and she was doing press interviews, as well as her own friends and her own family dealing with it. So, I felt slightly unsure – not to the point where I would ever want to question it – but I was conscious of the fact that I was unclear if Saima was a friend of mine, or if she was the subject for a podcast and, you know, that's always difficult to resolve in any circumstance. But in this circumstance, obviously, it's quite unique. So when she passed away, we found out via Facebook post or Instagram post, and that's how I became aware. Obviously, when people are close to you normally you would get a call or you would kind of be informed in a way but because of the situation of us being locked down in a pandemic and not seeing anybody anyway, I wasn't able to see Maleena when this happened. Normally, Maleena and I would have got together over a cup of coffee and talked about it and talked about how we felt but that wasn't viable at the time. Then it's this kind of Instagram world of finding weird stuff out via a mobile device rather than having a face-to-face conversation. Then Maleena and I had this kind of odd conversation about how we should reach out. People were even coming to us and giving their condolences and we were a bit like, we don't know if it's us that you should be giving condolences to. Then we thought: *how do we project our condolences?* I think that Maleena ended up sending [Saima's account] an Instagram DM, which felt a bit insane. But that was our only option, because we didn't have the phone number of her mum or her sister or her husband. I remember having a conversation with Maleena about whether we should ask about a memorial and do we need to inform the BBC? ... [In the end] they put screengrabs of her Instagram on their page. But it didn't feel like it was the right way of managing that. We should have had a better system in place of what to do when it happened, but we just never really got to that point.

Maleena: This is where it was difficult for me to navigate, because I had a slightly closer relationship to Saima. And I would say that we had become friends. I had met her sisters a few times. And as somebody who lost a sibling, I kind of immediately just wanted to reach out. So that's what I did. The thing is, going back to the point of finding things out digitally or on social media, we are now in a culture where that is how we get a lot of information, especially when you're talking about somebody that lived a lot of her life on social media. A lot of her community, a lot of the people that she was friends with, or were part of that community, were there in that space. And I sort of saw us as a bit of that too. But I think culturally it felt a bit insincere and insensitive [for us] to message her on DM and not give her [family] a call to say, 'Look, we're here if there's anything that you need', or have a space to go to, to sort of grieve together. That was I think the biggest challenge of not anticipating how this would play out in a pandemic – we had processed what it might be like to lose her, but not necessarily at this time. I think the thing that threw me off was, I couldn't connect to anybody, there was no connection, there was no opportunity to grieve her. I couldn't meet up and I couldn't reach out to anybody that was significant in her life. So the only option I had was to send a message to her and just hold space and say, 'Look, we're here if there's anything that you need', and [her sister] immediately got back to me and said thank you and explained that they had wanted to tell us, but they had so many people to tell because there were so many people in her life. I hadn't been expecting a phone call, but it was important, on a personal level, to reach out and ask if we could help in any way, to say that we'd love to share our condolences, and to ask if there was going to be a funeral because I'd spoken to Saima about it, we'd talked about her funeral. So I'd just assumed that we would be there to witness that journey. [But, of course, in the end it had to be] a small private family affair, because of the pandemic. So I think it was all a very, very unique situation and set of circumstances that made it a difficult experience.

Joining the Zoom [funeral] was weird, but also I'm glad that we were able to witness it. For that kind of closure, to a certain extent. I'm glad that I saw it, because I had imagined it with her. [And because] I'd talked about her funeral ... this was all the candid stuff we would talk about in between takes and during takes. So that was a resolution piece. It was really important to me that I witnessed that.

Faraz: I think the issue is that there is this kind of public grieving, which is what funerals allow you to do, and the opportunity to talk to people, etc. And then there is the kind of private grieving, and you know, they are two very, very separate things ... the same is true with family, sometimes you get invited to a funeral for a family member that you only met a couple of times, you should be there out of respect or obligation, but you don't have a particularly emotional connection with them.

This scenario was unique in the sense that, for myself, there was both. We had a situation where we had this private grieving due to being a pandemic, and us also having a unique relationship with her, because we had done this project with her. But then there's also this very, very public thing, which is people messaging us online saying, 'Oh, we heard about Saima and we're really sorry. And are you okay?'

Maleena: Yes, honestly, [at the time] I felt like that was a misplaced condolence. I wanted them to direct this towards her family ... [Now I realise that] the people that did reach out directly to us [was because] they had learned about her journey and who she was through us, therefore, we were that connection. And that's where they were able to share that feeling.

I was invited to talk about this project three times on advice podcasts, on an app and a bunch of different podcasts, really soon after Saima passed away. In fact, it was supposed to be the week that it happened. And obviously we didn't know it was going to happen. [So when it did], I just emailed the producer, and said I was just really sorry, I can't do this. I wanted it to be meaningful, and I didn't want to be over emotional. So I asked to push it back by a couple of weeks. [But then] I was able to sit very shortly after she died and talk about what that project was about and what it meant. And also try and honour her memory at the same time. And that happened a couple of times quite soon after it happened. Since then, I haven't talked about it. Then me and Faraz went to Masala Wala [Saima's café] about a month and a half ago, and we saw her sister, and I remember it was a [beautiful day] and the sky was really blue and Faraz was saying the sun is so bright, and it was like there was some real white magic in the air. Like it was the perfect time for us to come in. That was also a really important thing for me to do, going back into the space that we'd recorded in, paying respects to her sister in person, eating the food again, enjoying the space again, and talking about her.

I think the overriding thing for me was that we were very conscious about what we were doing at the time [with recording the podcast with Saima], going into it very consciously. We managed the reality of what would happen when she would leave us and that I always understood that this was so much more than just a podcast, this was an opportunity to create a legacy for her family, but also for us to have forever. You don't often get to immortalise someone in that way. And that makes it a different kind of loss.

Faraz: For me, there is a kind of significant difference between this and the losses that we've experienced elsewhere. In the sense that the conversation that we started to have with Saima was the fact that she was going to die, from the very second we met her ... So that kind of shock of finding out somebody's ill, we never had that, because we always came to her with that context in mind.

But while we were recording the podcast, Saima made us very, very aware that she may not make it to Christmas. So once we got over that time, and because we're in the pandemic, it took us much longer to make the podcast than we hoped it would do in the first place. Actually, we were almost grateful for that fact because we felt like we had more time with her than we expected to. That meant that the sense of grief wasn't the same, because it was like we had been given more, rather than had things taken away from us. Usually when you're dealing with grief, it's because things have been taken away from you. So I do think it's a different and unique experience ... I do feel grateful for both Saima allowing us to document her journey, what she was going through, and also for the fact that we got to spend more time with her than we expected to in the first instance ... I think that that's a gift rather than something that's been taken from us that we should mourn.

## GRIEF RITUALS

Maleena and Faraz powerfully shared their experiences of saying goodbye to Saima. That act of saying goodbye varies hugely around the world, but often takes the form of a funeral rite. However, funerals don't always offer the 'closure' we're looking for.

When I lost my dad, and our lounge turned into a wailing parlour, this too was a collective sharing of grief. But I didn't see it that way at the time. I saw it as this alien culture that had infiltrated my world.

This was my grandmother's way of grieving, as it fit with her cultural map of the world, and I had no idea how to grieve in this way, so I became very angry. I became angry because she was telling me how to grieve, and I realised then that nobody can tell you how to grieve. You can't copy and paste grief. You find your own way of dealing with it, and this is very personal to you.

When I discussed our different cultural rituals for grief and my grandmother's approach with cognitive neuroscientist Dr Lynda Shaw, she commented:

> There are three ways we can influence the brain. Our brain is influenced by our genetics; by our perceptions and beliefs, which is top-down processing; and by our environment and social upbringing. What your grandmother was doing was how she was culturally brought up. That is excellent. That would suit her and that is what we should actually do. I'll give you an example. My husband went to the funeral of a dear friend. He had to get back to work and he didn't have time to go to the wake. Because he didn't go through the ritual in our culture of going back and talking about the person – 'do you remember, when …' and all of those things – he didn't go through the ritual [in its entirety] and it had an adverse effect on him for days afterwards.

Elaine is a long-term client of mine and dance teacher. She very sadly lost both her parents during the COVID-19 pandemic and, due to the circumstances and restrictions in place at the time, the funeral arrangements were vastly different from 'normal'. When we spoke, she told me about how this impacted her ability to say goodbye, and how the differing circumstances of their illnesses and deaths changed her experience of grief.

# ELAINE'S STORY

Mum and dad's journey to their end, I suppose, was totally different for one than the other. With my mum, because she was diagnosed about eight years ago now with Alzheimer's, when she had that diagnosis, and a little bit before that, our relationship started to change. Although at that stage, she knew exactly who we were and everything. So, it was almost as though my journey started probably about eight years ago with sort of a living grief, but not really realising it until you get to the other stage that I'm at now. Because at that stage, when I was going through that living grief, it was a question of doing everything you possibly can, even from buying little gadgets and memory things, to try and keep what memories she had. But obviously that wasn't going to happen ...

She was looked after tremendously in the home, but the decline from her going into the home in 2018 was rapid. I think it was because of the changing circumstances, and she very quickly didn't really know who we were, because she got all mixed up with the staff. As much as I tried to give her memory books, and we did photograph albums and all that, and sometimes when we mentioned dancing, she'd come to life for a little bit, but you could gradually see the drift happening.

Christmas time was really hard because I like to make quite a bit of fuss about it for my parents. We went to the home and said 'Look, what you've got, and this is from Adam. And this is from all of us', and we made a big fuss about it. But by her last Christmas trying to open the presents, you know, just opening the paper was a non-starter really. So that sort of grief process is different, completely different from my dad's.

When Dad started getting ill, I was in and out of hospital, visiting him. Again, the same mechanism I used was to do practical things, food wise or getting in various gadgets to try and get him healthy again. That was okay, for a couple of weeks when he was out of hospital, and I was back home, back to sort of normality. Then usually two weeks down the line, I would get a call and it all happened again. That was the pattern for a long time really.

I suppose it was this time last year when things got quite bad. We were thinking about Christmas, wanting to go down and see mum in the home and then bringing dad back up here for Christmas, which we did. I brought him back two days before Christmas, and he was fine that one day, then the next day he ended up in hospital. His health just wasn't good. They sort of patched him up and shipped him back out again on Christmas Eve and looking back now, I'm thinking, *How did I do all that as well as Christmas party organising, all the Christmas presents and all the cards to be sent? And I don't know ...*

Eventually the doctors did say because his heart was so bad, it wasn't really clearing up all the infections and that's why they were coming back worse than ever ... By mid-January, it was getting worse and I stayed down there permanently and came back in February. I stayed down there for about three weeks in the end because we knew that that the end was happening. We got all the medical equipment delivered and everybody was marvellous, he had as much care as he could possibly have.

[I was anxious] having never really witnessed death before, which I hadn't. Only when my granddad died and then he was in hospital. And then I was a bit removed from it then because I was only 11. It wasn't my responsibility. I think that is the key. I now felt very responsible. And I stayed ... And I was there when he died. And I'm glad I was. It was just me and him ... [Initially] they wouldn't administer end of life care, even though we knew what was going to happen, because he wouldn't say that he was in pain. They couldn't do anything about it. And that was probably one of the hardest because he was fighting till the end. Eventually, he said, 'Yes, I am.' And they could take over and then administer pain relief, which virtually put him into a little bit of a deep sleep – he could still hear, and some people do come around. But most of the time, they just sleep and they're sort of very conscious of what's going on. About two or three hours before it happened, he had this horrible breathing, and I can hear it and smell it now. It's just one of those things that will never leave you. They call it the death rattle, apparently, it's just a horrible sort of noise. I do believe very much in fate, because on the Saturday he died, [my husband] and I went out and my nephew came up to sit with him for a couple of hours, we knew it was pretty bad, he'd been on the end of life for about three or four days. We went out for something to eat to just get away for an hour and came back. After we got back, his breathing was worse. The other weird thing really is he waited till I was there, I know he did. And it was Leap Day. So that sort of makes you have a bit of goosebumps, really. The fact I was there and the fact it was a leap day that comes only once in four years, that was really quite strange.

For me, I had to be there. The biggest fear every time I left when he was in the hospital, they said, 'Oh, well, you know, we're not sure he's going to pull through', because it was so bad. And each time he did, because he wanted to be at home, he didn't want to go in hospital, he didn't want to die there ... I was so grateful that all the times it could have happened previously, [in the end] he was where he wanted to be. Afterwards, every time I go in there, I can just see it. Our dog was with us a few times and he could just tell because they're so sensitive about the smell. I mean, obviously the smell wasn't there. But it's just the presence of that happening in that room. And I do think that. I don't know, but I think it will always be there in some respects.

I think the hardest thing as well when dad died, was that my mum was still alive but she didn't know her husband ever died. That was so weird. In some ways, it was a nice thing, because she never had to go through that grieving process.

Dad was religious. He enjoyed going to church, that was his one sort of get-out clause. 'Well, I've got to go to choir tonight, or I'm going to church', because that's a social thing as well, as much as anything else. And I think that for him was really good, because that got him through the tough times. So I knew that we needed to do a service at his church for him. But it wasn't just a service for dad, it was for them both ... Dad was in the Royal Artillery during the war ... and it was through contacts learning of dad's death that they organised the funeral and said, 'Would you like a bugler? To play the Last Post? We can get a regimental drape for the coffin.' It was a bit overwhelming really because you think, here is this person who lived until he was 96. One of his wishes was to visit the Cenotaph for Remembrance Day, a wish we were able to grant with help from his regiment in November 2019 and these circumstances led him to almost have a military funeral, this was incredible. We did his funeral on the Friday, before we went into lockdown on the Tuesday. If he'd have been there he'd have been so happy. And for myself, I was proud of what I'd achieved. Because over the time, I'd said, 'What's your favourite hymns', etc. ... and even the choir that he used to sing in [was there]. They asked to sing his favourite anthem, which was like, wow. And the vicar obviously knew him. We did pictures and a slideshow in the church, and particularly ones from the Cenotaph and family. So at the time, I thought, 'Gosh, the hardest thing is, do you then say to Mum, come on, you're going to dad's funeral.' She hadn't been out of the care home for two years. And we took the decision, that we weren't going to tell her and we weren't going to take her to the funeral. Somebody else might have said we should have done. But my feeling and [my sister's and the whole family's] feeling [was that we shouldn't] ... It would have been so stressful for her and even more so for us knowing what day it was, but Mum wouldn't understand. The kindest thing was to just not tell her.

The only thing that wasn't quite normal at that stage, this is only a few days before lockdown, a lot of the congregation, because they were older, were really quite worried about coming out. Some of them did because they wanted to come for my dad regardless. And I'm sure we'd have had a packed church if it was not for the pandemic that we didn't really know about at that stage.

The other nice thing was that the vicar came up to the house after [he died]. We had little chat, and he said some prayers. Then he said, 'Bring some things to the funeral that are meaningful'. So, I took their wedding photograph. And that was me saying that Mum is here. We took a few things and he said, 'When the choir is doing their anthem, just bring them up to the front, and that's what we did, we took his football scarf, which was always a favourite because there was always had a bit of a banter about it. So that was nice ... Then I suppose the only and the other nice thing that we did do, although it was only family afterwards, we went to the British Legion club.

After that, Mum's care home was in lockdown. And that was the hard thing because we just couldn't go and see her. So, we were still grieving dad. But then I was grieving mum as well. And even though she didn't know about Dad, I think it would have been a comfort to be able to go and see her at that stage. And I never saw her again.

She died on the 26th of April, of course, we were in the lockdown in the pandemic. The care home said she'd been bedridden for a little while and lost the use of her limbs a little bit, so couldn't walk much. But this was all part of the process. A couple of times before dad died, we went to see mum. And you know, she said a couple of things that she would only ever have said before she was ill which was really weird at the time and you think, *Wow, where did that come from?* Sometimes people can have moments when there is a bit of a flashback.

The care home rang me on the Thursday, and they were so lovely. They said that mum was deteriorating and asked for our permission to get the end of life care from the doctors in place. They said it's not that she's dying at the moment, she's okay. But they said, they were aware it was the weekend coming up and they didn't want to not have that for her if it was needed. So obviously, we gave our permission, that wasn't a problem. Then they rang me on the Saturday morning, and it was a lovely day. We were sat in the garden, when we took the call and they said, 'Your mum's getting towards end of life, we've started to administer the end of life care'. So, my immediate question was can I see her? And they said, 'It's totally up to you. We're not going to say, "No, you can't." We have got Coronavirus in the home. You do have to have full PPE. And obviously there is a big risk, then you do have to isolate for 14 days afterwards.' So I said okay, 'I'll think about it and get back in touch.' And my husband said, 'If you want to go, we'll go now, it's not a problem.' I still don't know what the right decision is and whether I would have changed my mind. I think I might have done, I might have taken that risk and gone. The big thing afterwards was that isolating would have been a problem. But I couldn't isolate at home, because my husband is a front-line worker. So, it wasn't practical to come home and isolate. I could have isolated at my mum and dad's but for my own mental health, I couldn't cope with looking at the space where dad had just died a month before. I just thought I've got to be a bit practical here. That's not going to do me any good. So, I took the decision I wasn't going to go. Because it wasn't as though she knew me. It wasn't like dad, who knew I was there, and I could hold his hand and talk to him, and I knew he could hear me. Mum could probably hear me [but whether she'd know it was me or a member of staff it was hard to tell]. We had a few incidents earlier that led me to believe that she may know we were there. My nephew because he'd been out working anyway, took the decision, to go to say goodbye for the family. He said it was funny because although she had her eyes closed, he talked about everybody in the family, bless him how he did that I don't know, he mentioned everybody's names to her, and he said she kept flicking her eyes when names were mentioned. And he said I mentioned your name, and she opened her eyes, whether she was opening her eyes to look to see if I was there, I don't know. You can read a lot into these things. And sometimes you can ask yourself questions and keep going over them. Then on that Saturday afternoon I thought to myself, *Right, I'm going to go down*. I'd almost sort of said to myself, I could go down on my own. And then if need be, I'll come back home. My husband can almost move into his mum or dad's or something. And I'll stay here. So, I'd almost made that decision. And then she died in the early hours of Sunday morning before I got there. A member of staff was with her at the end, and they said she had Coronavirus symptoms. She never had a test and she had Alzheimer's for eight years. Yet, on her death certificate it's got Coronavirus with underlying Alzheimer's, but it was the Alzheimer's that killed her, it's got to have been and I think that's the hard thing to swallow now. If she had had that for that length of time, I don't know, a chest infection would have killed her anyway, the same as my dad because they just weren't strong enough to fight back from that.

I don't know, part of me thinks, the logical part of the brain says, Thank goodness for that, because she didn't have a life. She was sat in the bed all the time. She didn't know anybody. She'd stopped eating a few weeks beforehand, she was on a liquid diet. You know, the whole scenario that you think of, this is just cruel, you wouldn't want anybody to suffer like it. And to see her, it was really cruel. So in the end, logically your brain says, *Yeah, you know, it was the best thing for her.* The other part of my brain says, *I just needed to see her that last time.*

[With Dad] I took a photograph every day for the last three days, even the day he died, in which you can see that he was starting to suffer. I had a little video that I haven't shown anyone because I know that it's upsetting. But I took a little video when dad was in bed, the weekend my son came down before he died of them singing a hymn together, because my son was sat there saying, 'Right which hymn shall we have granddad, oh, let's play this one'. So, I have them singing together, just a few minutes of a clip of it. It's so special to have dad's voice. And I think the saddest thing, thinking about it now, I've got a couple of videos of mum, but I wish I'd have done more of that, to keep more of the memory of the person that she was, the strong person that she was, not the declining person that I saw towards the end.

When dad died, we had the undertakers arrive, we were able to discuss what arrangements we would like and we gave them his favourite suit, and he had a replica set of medals, a small set that we actually put in the coffin, we made sure he had a photograph of him and mum on their 60th wedding anniversary and photograph of the family. And so we were able to do all of those little things that make you feel a bit better. In a weird sense. We were able to go to the chapel of rest and see him there. Yes, you can tell His Spirit had gone, obviously. But he looked nice and peaceful. That was good.

When mum died, I couldn't see her. You couldn't dress her in anything. You couldn't say, 'Oh, well, let's put her in something that she was happy in' or, you know, 'let's put a picture of me dancing' – you couldn't do anything to make yourself feel better. No, I just had to go with it. And I think that was the hardest thing. I was more upset at mum's funeral, than I was at dad's. I was pretty bad at dad's, but you knew you were able to produce something nice for his end and mum's, well, we didn't see her until we got to the crematorium.

[At the time,] different counties had different regulations to do with funerals, so dad's was fine because that was before lockdown, but mum's was in lockdown. We were allowed ten people. We weren't allowed to have flowers and she loves flowers. We weren't allowed to have all the little personal things that you would have done. We couldn't do any of those. The funeral director did say, 'If you buy your own flowers, you can just put something on the coffin when you go up for the committal.' So, I bought a dozen red roses. Then when we played a certain song, we just had to sit socially distanced from each other. We were all sort of separate, which was really quite cold. I still did a nice eulogy. Because it's something that I needed to do as well, because it was the one thing I could do was the order of service. I was upset, because it was out of my control, with dad's funeral, I had a bit of control. So, it felt really cold. But it was a lovely service. We had 'Somewhere Over the Rainbow', because I like to think when I see a rainbow that she's gone over the rainbow to the sunshine to dad at the other end. So that was lovely.

There should have been a live web link but it failed which was rather upsetting. We did however get a recording afterwards.

I think the whole pandemic was hard because we couldn't have a wake, only a sandwich in the back of the car.

I think having Alzheimer's is hard, but also losing your dad and not being able to then reach out to your mum at a time that you should be able to is even harder. But what a nice thing for her to never know that her husband died. I think to myself, well, that's really good. Neither of them were afraid of death. I think dad fought so hard because he never wanted to leave mum. Yet strangely, mum had already left dad in her mind, but she didn't have a choice in it.

That's why I wanted to [do this interview and tell my story], I knew this would be hard. But I think it's really important, because I don't know how you deal with grief ... because it is a living grief that I've been dealing with for eight years. And although this year has been the worst, but you don't realise that until the time comes and you're able to process it a little bit more ... Because at the end of the day, there's only you that can actually deal with it and help yourself. Since this time very sadly we lost my father in law and five months after my mother in law who took her own life. Our grief is ongoing and dealing with this is a daily challenge but life goes on and we have to live that life.

Elaine's story will sadly be only too familiar to many people who have lost loved ones, and especially to those who have lost them during the pandemic when the 'normal' grief rituals have no longer been available to us. What it does remind us of is that we value choice, and that, although we may not be able to choose how or when we die, we can – if we overcome our societal fear of discussing death – make choices while we are alive that will not only help us but also help our loved ones on that grief journey.

Siân Storey is a funeral celebrant, and she runs her own Death Café, where people go to discuss death while enjoying tea and cake. She started her career as an intensive care unit nurse and spent a lot of time with patients who were dying and felt enormous gratitude to be able to sit and speak with the dying person, even as they were unresponsive. Siân tells me that even when dying, the auditory part of the brain is still stimulated. Dying people, even in a coma, can hear everything that is going on, and even though they may not be able to respond, they can still hear you.

Siân also loved to be there for the families of the dying person, and she developed a passion for making death a beautiful experience, and now her mission is to help people do death well. By being a funeral celebrant and hosting the regular Death Cafés, she is able to hold the space for death with compassion and beauty. Siân describes her job as joyful. She loves what she does, because she sees it as an absolute honour to celebrate the person who has died in a meaningful way.

215

The Death Café movement is very encouraging, and the rising popularity of them as a sharing space to discuss death and dying is heartening. It shows that we are finding spaces to open up about grief and change the conversation.

I've spent a lot of time in this book talking about people who go to great lengths to avoid thinking, discussing and contemplating death, and equally avoiding or trying to suppress grief. I asked Siân why this was, and she tells me that, believe it or not, not everyone shies away from death. In fact, there are some Swiss euthanasia clinics that offer a 'How to end your life' guide, and some where you can choose your death day and check yourself into the clinic, and that will be that. To qualify, you have to be over 50. You don't need to have anything wrong with you or be dying anyway – you have free agency to decide when you want to end your life.

For a lot of people, this is hugely important, to be able to be fully in control over when they leave this planet. This is certainly an interesting topic of discussion, and there are now several places around the world that will assist you in dying. I think the right term for this is Voluntary Assisted Dying. It is illegal in the UK, but it does go on. It's certainly a discussion that divides the room and people will react strongly either way. I prefer to stand in the middle of the room, where I can really see both sides and that feels good for me.

## SAYING GOODBYE

Andy Chaleff, Californian-raised author of *The Last Letter* and *The Wounded Healer*, tells me that while growing up he had always been more sensitive than the people around him. He didn't see himself fitting in with the status quo. He was the 'different' kid.

He likes to call out discomfort. He describes himself as being attracted to things on the outskirts of most people's comfort zones, which is why he thought about death so much as a teenager and gave himself night-time panic attacks, from which he would wake up hyperventilating. He had a fear of non-existence. He never told anyone about this fear and kept it to himself. His biggest worry was that death was possible.

He told me a bit about his background, describing his dad as emotionally absent and abusive. He would avoid contact and interaction with his father. Instead, he would seek emotional support and love from his mother. Andy and his mother were incredibly close. Andy saw his mum as his best friend. She was his rock and always knew what to say to make him feel better.

When Andy left home and went to study his degree at UC Irvine, he missed his mother so much – he realised he was mapping his fear of death around being away from home, and the homesickness was effectively giving him symptoms of grief.

Andy, being Andy, decided to take a deeper dive into his own discomfort and confront his fear of death head on by taking a course offered on his degree pathway that was entitled Sociology of Death.

This course raised a pivotal question: if you were going to die tomorrow, and you could do one thing today before you died, what would it be?

Perhaps it was a rhetorical question, or perhaps it was a question to spark an idea, but it certainly had an impact on Andy.

He decided to write his mother a letter. A letter written from his heart to express his thanks to her for everything she had done for him. A letter to tell her how he felt about her and their relationship. He poured his honesty on to the paper, and immediately after he had finished writing, he went to the post box to send it so she would receive it as soon as possible.

A few days later, when he had come back to his room from the running track, his mum had left an answer-machine message telling him she got the letter and how touched she was by it.

Andy describes a deep sense of gratitude that she got the letter, and she felt strongly that this was a gift from him to her.

The next morning was Saturday, and Andy was back at the running track, training; he looked up and saw a figure walking towards him, and as the figure got closer, he recognised his older brother. Immediately, he knew something was very wrong.

Andy's brother had come to tell him that their mother had been hit and killed in a road-traffic accident. A drunk driver had killed their mother.

Grief filled Andy's system with numbness, depression, shock and disbelief. He retreated into his shell and stayed there for a long time. It was a year later when he felt strong and brave enough to emerge. He asked for help and began his own journey into self-development. Ultimately, skipping forward many years, his grief drove him into coaching and mentoring, and from this experience Andy took an incredible road trip, where he would go on to help many others with their relationships with death, pain, grief and loss.

Andy suggests that you can do this letter-writing exercise:

Write down 'Dear ...' and whatever comes next.

It could be 'Dear Dad' or 'Dear Pain' or 'Dear Mum', 'Dear God', 'Dear Exhaustion'.

There is no rule about what you do with this letter. You have full choice whether to keep it, burn it, send it or rip it up. Whatever feels right to do in that moment. The letter is a flow and a stream of con-sciousness: it doesn't need to make sense – there are no restrictions. It is a free, non-thinking exercise. Just write it all out. Don't edit it. There is no wrong. It's all right.

If you don't know where to start, you can start with this: 'Dear Grief, I don't know where to start. I don't know where to start, I don't know where to start ...'

Keep writing and something will come, and even if it doesn't, that's okay too. This is a powerful exercise, because your subcon-scious will know what it needs to release. Once you have the letter, what you do with it next is your choice. This is your process. You have an opportunity to take control of this letter. It's your journey and you can take it in many different directions. You have agency. There is no right way or wrong way to heal. There is your way, and that is all right.

Death can really throw up some challenging stuff can't it, and then, with that discomfort, perhaps we can say the wrong thing, or

DEATH: BACK HERE AGAIN!

not know what to say? Maybe we can stumble over our words, or just sit in uncomfortable silence. Silence is an interesting concept. Some people hate it, and others crave it.

Andy Chaleff describes being in an uncomfortable silence one time, and then vocalising how the silence was uncomfortable as a form of release. I asked Siân what she thought about silence, and she told me that silence is another form of communication. No words are sometimes better than lots of unhelpful words.

Siân, Andy and I share the same belief that often the most commonly used 'go-to five phrases' of support are actually very unhelpful. Please do not say any of these phrases to someone who is grieving:

1  They have gone to a better place
2  Time is a healer
3  They are watching over you
4  I know exactly how you feel
5  They had a good life

If you don't know what to say, holding the space for someone in silence is really supportive. Say 'I don't know what to say, but I am here for you'. Some people are very uncomfortable addressing death, and will avoid speaking about it, but these people still need to be held in a safe space. Once we become comfortable with death, we also become comfortable not speaking about death as well.

There is an urgency to be there for the person who is grieving. Siân also feels that it's her duty to bring an urgency of celebration into the funeral. To be able to laugh as well as cry, remember the good times as well as the bad. To be able to send that person off in a way that, if they were there, even they would be smiling.

The funeral is the last goodbye in a formal setting, but of course we can have our own private rituals and do our own thing to make peace with our loss. This personalises our goodbye and can give us a powerful sense of personal closure.

Andy Chaleff learnt that the concept of death creates an urgency to live life well. He describes life so poetically: 'We are born and we die. And then, in the middle, there is this thing called life.'

As he was telling me this, in my mind I had conjured up a vivid image of a hugely filled sandwich. The two layers of bread, also known as birth and death, were holding the sandwich filling of our life together.

His point is that life needs to have the birth thing and the death thing to hold it all together – like a container – so it seems ridiculous that we accept birth and spend our lives trying to avoid death. What if the two slices of bread that hold our sandwich filler together are held with equal reverence?

He tells me that non-urgent life keeps us, as a society, reasonably unhappy and we all learn to accept this level of discomfort as normal. This level of discomfort is also known as chronic underlying stress, which has profound impacts on our bodies, as Dr Shaw described so powerfully at the beginning of this book. Andy says that this type of underlying stress comes from the head and seeps into our physiology. To combat this stress, we need to find a sense of urgency to deal with it, and this sense comes from listening carefully to the calling of the heart. Stress from the head sets us into wheel-spin mode. Urgency from the heart gives us a road to drive on. Once we learn to use the heart to guide us, we develop a trust in something that needs to happen and it doesn't need a risk analysis – it just happens.

Life is too short not to act. There's nothing to gain by being comfortable. It stops growth. Discomfort becomes normal. We learn to accept discomfort and accept pain, and learn to live with stress, until there comes a point where we begin to crack, or we become so fragile that a tiny knock will shatter us into fragments.

Interestingly, Andy's letter-writing experience is one that Dr Shaw also recommends to her clients as a way of being able to allow the piece of paper to hold the thoughts, so that you can let them go. She suggests not reading the letter again, but rather putting it in an envelope, addressing it and then ceremoniously burning it. In this way, it gives space to the grief, honours it and initiates a healing process.

We don't move on from grief; as we have seen, it's part of the very fabric of all our lives, from birth through to the end. Our experiences of it can be varied and feel complex, in part because we often don't name and claim it for what it is, and I hope that this book will help you do just that, as in living your grief you can use its power to get you where you want to be.

Throughout this book I have shared many client experiences and stories, and interviewed many people with varying voices and ways of thinking; through talking to all of these people, I have discovered that, although we as human beings are different in so many ways, we do have this commonality. Ultimately, all human beings want to be heard, listened to, understood, managed well, accepted and acknowledged in some way, and we want to live a purposeful existence and feel needed. So, with this in mind, I've come up with my own version of Maslow's Hierarchy of Needs:

**H**eard
**U**nderstood
**M**anaged
**A**ccepted
**N**eeded

Throughout this book, we've explored many words around grief, but ultimately I love the way Andy Chaleff describes grief. It is a simple definition and distils this book beautifully. According to Andy, grief is dot dot dot. Just like a fingerprint, everybody has their own unique experience with grief.

After all, grief is the stuff of life, and without life, grief cannot exist, and without grief, life cannot be sustained.

I hope that within these pages I have shown you how we can convert grief into fuel, and how we can use this fuel to power our life.

*Grief is love repackaged* – and all human beings, animals and our beautiful planet could definitely do with much more love right now.

# HELPLINES AND SUPPORT RESOURCES

Throughout this book, we've touched on a number of complex, difficult and sensitive issues, and it's always important to seek professional help and support when dealing with loss, anxiety, trauma and concerns around your mental health or the mental health of others. The following is a short list, but some good places to start if you are based in the UK:

**BBC Headroom:** the BBC have brought together a mental health toolkit, particularly to offer support to everyone impacted by the pandemic; you can find this by searching for BBC Headroom or BBC mental health toolkit

**Cruse Bereavement Care:** 0808 808 1677; cruse.org.uk

**Mental Health Foundation:** mentalhealth.org.uk

**Mind:** 0300 123 3393; mind.org.uk

**The Samaritans:** 116 123; Samaritans.org

**Self-Injury Support:** 0808 800 8088; selfinjurysupport.co.uk

**YoungMinds:** 0808 802 5544; youngminds.org.uk

Following on from reading this book, you may find it useful to explore Dipti's free seven-day course for coping with loss, which can be accessed at: diptitait.com/welcome/planet-grief/

# SELECT BIBLIOGRAPHY

Agarwal, Pragya, *Sway: Unravelling Conscious Bias* (Bloomsbury Sigma, 2020)

Ashdown, Annie, *The Confidence Factor: The Seven Secrets of Successful People* (Crimson Publishing, 2013)

Bolte Taylor, Dr Jill, *My Stroke of Insight: A Brain Scientist's Personal Journey* (Hodder, 2009)

Chaleff, Andy, *The Last Letter: Embracing Pain to Create a Meaningful Life* (Koehler Books, 2018)

Crofts, Neil, *Authentic: How to Make a Living by Being Yourself* (Capstone, 2003)

Dilts, Robert, Hallbom, Tim & Smith, Suzi, *Beliefs: Pathways to Health and Wellbeing* (Crown House Publishing, 2012)

Forte, Sandro, *Dare to Be Different* (Professional Sales and Marketing Pte Ltd, 2015)

Gilligan, Stephen & Dilts, Robert, *The Hero's Journey: A Voyage of Self-Discovery* (Crown House Publishing Ltd, 2016)

McDermott, Ian, *Boost Your Confidence with NLP: Simple Techniques for a More Confident and Successful You* (Piatkus, 2010)

Peters, Prof. Steve, *The Chimp Paradox: The Mind Management Program for Confidence, Success and Happiness* (Vermilion, 2012)

Power, Penny, *Business is Personal: Be the Leader of your Life and Business* (Panoma Press, 2019)

Shaw, Dr Lynda, *Your Brain is Boss: Using Mind Power to Develop Influence, Creativity and Work Satisfaction* (SRA Books, 2017)

Shetty, Jay, *Think Like a Monk: Train Your Mind for Peace and Purpose Every Day* (Thorsons, 2020)

Storr, Farrah, *The Discomfort Zone: How to Get What You Want by Living Fearlessly* (Piatkus, 2018)

Tait, Dipti, *Good Grief: A Companion to Change and Loss* (Balboa Press, 2016)

Tripp, James, *Hypnosis without Trance: How Hypnosis Really Works* (Real Magic Media, 2021)